T0137069

# Description Logics in Multimedia Reasoning

Leslie F. Sikos

# Description Logics in Multimedia Reasoning

 Springer

Leslie F. Sikos
Centre for Knowledge & Interaction Technologies
Flinders University
Adelaide
Australia

ISBN 978-3-319-85308-6        ISBN 978-3-319-54066-5   (eBook)
DOI 10.1007/978-3-319-54066-5

Printed on acid-free paper

This Springer imprint is published by Springer Nature
The registered company is Springer International Publishing AG
The registered company address is: Gewerbestrasse 11, 6330 Cham, Switzerland

# Preface

The immense and constantly growing number of videos urges efficient automated processing mechanisms for multimedia contents, which is a real challenge due to the huge semantic gap between what computers can automatically interpret from audio and video signals and what humans can comprehend based on cognition, knowledge, and experience. Low-level features, which correspond to local and global characteristics of audio and video signals, and low-level feature aggregates and statistics, such as various histograms based on low-level features, can be represented by low-level feature descriptors. These automatically extractable descriptors, such as dominant color and motion trajectory, are suitable for a limited range of applications only (e.g., machine learning-based classification) and are not connected directly to sophisticated human-interpretable concepts, such as concepts depicted in a video, which can be described using high-level descriptors only.

To narrow the semantic gap, feature extraction and analysis can be complemented by machine-interpretable background knowledge formalized with description logics (DL) and implemented in ontology languages, in particular the Web Ontology Language (OWL).

While many knowledge representations employ general-purpose DL, multimedia descriptions, such as the description of moving objects in videos, may utilize spatial, temporal, and fuzzy DL as well. This enables the representation of topological relationships between spatial objects, the representation of the process state and state sequences, quantitative and qualitative spatial reasoning, handling imprecision and uncertainty, and selecting the most probable interpretation of a scene.

The depicted concepts and their relationships are usually described in the Resource Description Framework (RDF), which can express machine-readable statements in the form of subject-predicate-object triples (RDF triples), e.g., scene-depicts-person. The formal definition of the depicted concepts and relationships are derived from controlled vocabularies, ontologies, common sense knowledge bases, and Linked Open Data (LOD) datasets. The regions of interest (ROIs) can be annotated using spatial DL formalisms and media fragment identifiers. The temporal annotation of actions and video events can be performed using temporal

DL and rule-based mechanisms. The fusion of these annotations, including descriptors of different modalities, is suitable even for the machine-interpretable spatio-temporal annotation of complex video scenes.

Based on these structured annotations, various inference tasks can be performed to enable the automated interpretation of images, 3D models, audio contents, and video scenes. For example, video frame interpretation can be performed via abductive reasoning and video event recognition via reasoning over temporal DL axioms. The structured annotation of multimedia contents can be efficiently queried manually and programmatically using the very powerful query language SPARQL, although high-level concept mapping usually requires human supervision and judgment, and automatic annotation options need more research. Research results for high-level concept mapping in constrained videos, such as medical, news, and sport videos, are already promising. The structured description of multimedia contents and the semantic enrichment of multimedia metadata can be applied in video understanding, content-based video indexing and retrieval, automated subtitle generation, clinical decision support, and automated music and movie recommendation engines.

High-level concept mapping relies on formal concept definitions provided by RDFS (RDF Schema) vocabularies and OWL ontologies. However, it is a common and bad practice to create OWL ontologies using a tree structure of concept hierarchies visualized by the graphical user interface of ontology editors, and in particular that of Protégé, without formal grounding in DL.

Without logical underpinning, the computational properties of ontologies remain unclear, and reasoning over RDF statements leveraging formal definitions of ontologies that lack formal grounding might be not decidable. Beyond decidability, a crucial design principle in ontology engineering is to establish favorable trade-offs between expressivity and scalability, and when needed maximize expressivity. However, the higher the DL expressivity, the higher the reasoning complexity. Since the best balance between language expressivity and reasoning complexity depends on the intended application, a variety of DL have been developed, supporting different sets of mathematical constructors.

The level of semantic representation and logical formalization, together with the knowledge base size, presence or absence of instances, and the capabilities of the reasoner, determines what kind of reasoning is feasible. The reasoning tasks utilize different sets of reasoning rules that provide semantics for RDF and RDFS vocabularies, RDFS datatypes, and OWL ontologies, to infer new statements, check knowledge base consistency, determine concept satisfiability, calculate the subsumption hierarchy, and perform instance checking.

The formal knowledge representation of, and reasoning over, depicted objects and events can be used in high-level scene interpretation. Application areas include, but are not limited to, classification, video surveillance, intelligent video analysis, moving object detection and tracking, human intention detection, real-time activity monitoring, next-generation multimedia indexing, and content-based multimedia retrieval.

This book introduces researchers to multimedia semantics by providing an in-depth review of state-of-the-art standards, technologies, ontologies, and software tools. It draws attention to the importance of formal grounding in the knowledge representation of multimedia objects, and the potential of multimedia reasoning in intelligent multimedia applications. It presents both theoretical discussions and multimedia ontology engineering best practices. In this book, the reader familiar with mathematical logic, Internet, and multimedia fundamentals can learn to develop formally grounded multimedia ontologies and map concept definitions to high-level descriptors. The core reasoning tasks, reasoning algorithms, and industry-leading reasoners are also presented, and scene interpretation via reasoning demonstrated. This book aims to illustrate how to use DL-based formalisms to their full potential in the creation, indexing, and reuse of multimedia semantics.

Adelaide, Australia                                                                                      Leslie F. Sikos

# Contents

# About the Author

**Leslie F. Sikos, Ph.D.**, is a researcher at Flinders University, Australia, specializing in knowledge representation of multimedia resources, multimedia ontology engineering, and automated video scene interpretation via spatiotemporal reasoning and information fusion. He has worked in both academia and the industry, thereby acquiring hands-on skills in Semantic Web technologies, image processing, video authoring, CGI, and 3D modeling. Dr. Sikos is an experienced author with a strong textbook publishing background, who regularly works with the world's leading publishing houses. He has developed two of the most expressive multimedia ontologies to date, which are formally grounded in description logics and complemented by rulesets. Dr. Sikos is a recognized expert in the standardization of next-generation video indexing techniques that leverage semantic annotation at different levels of granularity, including entire videos, video shots, video segments, video frames, and regions of interest of video frames, and mapping depicted concepts to Linked Data. Inspired by the creation and exploitation of rich LOD datasets, Dr. Sikos actively contributes to the development of open standards and open data repositories. For more information, visit https://www.lesliesikos.com.

# Chapter 1
# Multimedia Semantics

In recent years, the production, storage, and sharing of the exponentially increasing number of multimedia resources became simple, owing to lower hardware costs, new web standards, and free hosting options on social media and video sharing portals. However, the contents of multimedia resources are, for the most part, meaningless to software agents, preventing automated processing, which is very much desired not only in multimedia retrieval, but also in machine vision. There are multiple approaches to address this issue, such as the machine-readable annotation of the depicted concepts, the formal description of scenes, and machine learning with pretrained classifiers, the latter of which is the primary means of automated structured multimedia annotation.

## 1.1 Rationale

So far, the majority of multimedia semantics research has focused on image understanding, and to a far lesser extent on audio and video semantics. Since the term *multimedia* refers to various combinations of two or more content forms, including text, audio, image, animation, video, and interactive content, images alone are not multimedia contents, but can be components of multimedia contents. Yet, the literature often uses the term multimedia to techniques and tools that are limited to capturing image semantics.

Inherently, video interpretation is far more complex than image understanding. Knowledge acquisition in video content analysis involves the extraction of spatial and temporal information, which can be used for a wide range of applications including, but not limited to, face recognition, object tracking, dynamic masking, tamper detection, abandoned luggage detection, automated number plate recognition, and lane departure warning. Without context and *semantics* (meaning), however, even basic tasks are limited or infeasible. For example, an attack cannot be differentiated from self-defense or training without seeing the preceding events. As

© Springer International Publishing AG 2017
L.F. Sikos, *Description Logics in Multimedia Reasoning*,
DOI 10.1007/978-3-319-54066-5_1

a consequence, video event identification alone is insufficient for classification. Automated scene interpretation and video understanding rely on the formal representation of human knowledge, which is suitable for automated subtitle generation, intelligent medical decision support, and so on. While computers can be trained to recognize features based on signal processing, they cannot interpret sophisticated visual contents without additional semantics (see Table 1.1).

**Table 1.1** Major differences between human understanding and computer interpretation of video scenes

| Humans | Computers |
| --- | --- |
| *Intelligence* | |
| Real-time understanding is straightforward in most cases, although movies with a complex plot may need to be watched again to be fully understood | Overwhelming amount of information to process; algorithms and methods are often insufficient, making video understanding infeasible even without time constraints, yet alone in near-real time or real time |
| Context is understood from plot, title, events, genre, etc. | Potential interpretations are extremely confusing; metadata, if available, can be combined with concepts mapped to common sense knowledge bases or ontologies |
| Visual content is understood (even without colors) | Automatically extractable features and their statistics convey no information about the actual visual content (nothing is self-explanatory) |
| Years or decades of life experience and learning make it possible to recognize virtually anything | Training from a few hundred or thousand clips provides a very limited recognition capability |
| General understanding of how the universe works (e.g., common sense, naïve physics) | Only tiny, isolated representations of the world are formalized, therefore unconstrained video scenes cannot be processed efficiently |
| Understanding of human behavior enables prediction | Only fractions of human behavior are encoded, so software agents cannot expect upcoming movements |
| The human mind and eyes are adaptive and recognize persons or objects moving to or in darkness, or in noisy or less detailed recordings (e.g., old VHS video, small-resolution video) | If the noise-signal ratio falls below a threshold, algorithms perform poorly |
| *Spatial information* | |
| 3D projected to 2D can be interpreted by stereovision: planes of graphical projection with multiple vanishing points are understood, which enables perspective viewing | Most videos have no depth information, although proprietary and standardized 3D recording and playback mechanisms are available; RGB-D and Kinect depth sensors can provide depth information |
| 3D objects are recognized from most angles | Training provides information for particular viewing angles only—recognition from different viewpoints is problematic; scale-/rotation-invariant features are used for object tracking in videos |
| Partially covered objects and persons are relatively easily recognized | Occlusion is problematic |

(continued)

**Table 1.1** (continued)

| Humans | Computers |
|---|---|
| *Temporal information* | |
| Continuity; events and happenings are understood even in the case of nonlinear narratives with extensive flashbacks and flashforwards (although movies with a very complex plot might be watched again to be fully understood) | Very few mechanisms for complex event detection; videos are usually compressed using lossy compression, therefore only certain frames can be used; no information can be obtained on complex events from signal processing |
| *Information fusion* | |
| Seamless/straightforward understanding of simultaneous multimodal information playback (e.g., video with audio and subtitle(s), hypervideo) | Information fusion is desired, which needs more research |
| Audio channel is understood relatively easily and conveys additional information | Algorithms for detecting distinct noises (e.g., gunshot, screaming) are available; complex audio analysis is a challenge |
| Subtitles and closed captions can be read by most humans and convey additional information | Text-based, timestamped subtitle files can be processed very efficiently; however, incorporating the obtained information into higher-level video understanding is still challenging |

## 1.2   Feature Extraction and Feature Statistics for Classification

Multimedia features extracted from media resources can be converted into numerical or symbolic form, which enables the automated processing of core characteristics. Low-level features, which capture the perceptual saliency of media signals, include visual features (e.g., color, texture, shape), audio features (e.g., loudness, pitch, timbre), and text features (e.g., speaking rate and pause length calculated by processing closed captions). Combining the results of video, audio, and subtitle analysis often provides complementary information. Production and licensing metadata, when available, can be used for aggregated semantic enrichment of media resources.

A wide range of well-established algorithms exists for automatically extracting low-level video features, as, for example, fast color quantization to extract the dominant colors [1] or Gabor filter banks to extract homogeneous texture descriptors [2]. Some of these features, such as motion vectors, are employed by video compression algorithms of state-of-the-art video codecs, such as H.264/MPEG-4 AVC and H.265/HEVC, and by video analysis.

The state-of-the-art video classification approaches exploit sparse local keypoint features, i.e., salient patches that contain rich local information about an image or a video frame. They apply the *bag of visual words* (BoVW) model using local aggregated visual descriptors, typically histogram of oriented gradients (HOG), histogram of optical flow (HOF), or motion boundary histograms (MBH).

Histograms are based on accumulative statistics that are not affected by small local changes of the content and are invariant to common transformations, such as signal scaling or coordinate shift. Well-established algorithms utilizing such descriptors are efficient in classification, video clip matching, and object recognition, but not necessarily in video understanding, particularly when it comes to scene interpretation and knowledge discovery. The bag-of-words model applies a visual vocabulary generated by grouping similar keypoints into a large number of clusters and handling each cluster as a visual word. A histogram of visual words can be constructed by mapping the keypoints back into the vocabulary, which provides the feature clue for multimedia indexing and classification.

## 1.3   Machine Learning for Multimedia Understanding

Low-level multimedia descriptors are typically fed into a recognition system powered by supervised learning, such as SVM classifiers (support vector machines), which look for an optimal hyperplane to find the most probable interpretation, such as via a Lagrangian optimization problem [3] (see Eq. 1.1):

$$\min_{\beta, \beta_0} L(\beta) = \frac{1}{2} \|\beta\|^2 \quad \text{s.t.} y_i \left( \beta^{\mathrm{T}} x_i + \beta_0 \right) \geqslant 1 \text{ for } \forall i \qquad (1.1)$$

where $\beta^{\mathrm{T}} x_i + \beta_0$ represents a hyperplane, $\beta$ the weight vector of the optimal hyperplane, $\beta_0$ the bias of the optimal hyperplane, $y_i$ the labels of the training examples, and $x_i$ the training examples. Since there is an infinite number of potential representations of the optimal hyperplane, there is a convention to choose the one where $\beta^{\mathrm{T}} x_i + \beta_0 = 1$. Once a classifier is trained on images depicting the object of interest (positive examples) and on images that do not (negative examples), it can make decisions regarding the probability of object match in other images (i.e., object recognition). Complex events of unconstrained real-world videos can also be efficiently detected and modeled using an intermediate level of semantic representation using support vector machines [4].

Bayesian networks are suitable for content-based semantic image understanding by integrating low-level features and high-level semantics [5]. By using a set of images for training to derive simple statistics for the conditional probabilities, Bayesian networks usually provide more relevant concepts than discriminant-based systems, such as neural networks. Papadopoulos et al. proposed a machine learning approach to image classification, which combines global image classification and local, region-based image classification through information fusion [6].

Other machine learning models used for multimedia semantics include hidden Markov models (HMM), k-nearest neighbor (kNN), Gaussian mixture models (GMM), logistic regression, and Adaboost.

## 1.4 Object Detection and Recognition

The research interest for machine learning applications in object recognition covers still images, image sequences [7], and videos, in which spatiotemporal data needs to be taken into account for machine interpretation. There are advanced algorithms for video content analysis, such as the Viola-Jones and Lienhart-Maydt object detection algorithms [8, 9], as well as the SIFT [10], SURF [11], and ORB [12] keypoint detection algorithms. The corresponding descriptors can be used as positive and negative examples in machine learning, such as support vector machines and Bayesian networks, for keyframe analysis and, to a lesser extent, video scene understanding. Beyond the general object detection algorithms, there are algorithms specifically designed for human recognition. Persons can be recognized by, among others, shape, geometric features such as face [13], and behavioral patterns such as gait [14].

Based on the detected or recognized objects, machine learning can be utilized via *training*, which relies on a set of training samples (*training dataset*). For example, by creating a training dataset containing cropped, scaled, and eye-aligned images about different facial expressions of a person, the face of the person can be automatically recognized in videos [15]. The training optionally includes a set of responses corresponding to the samples and/or a mask of missing measurements. For classification, weight values might be given to the various classes of a dataset. Weight values given to each training sample can be used when improving the dataset based on accuracy feedback.

## 1.5 Spatiotemporal Data Extraction for Video Event Recognition

Automated video event recognition is amongst the most important goals of many video-based intelligent systems ranging from video surveillance to content-based video retrieval. It identifies and localizes video events characterized by spatiotemporal visual patterns of happenings over time, including object movements, trajectories, acceleration or deceleration, and behavior with respect to time constrains and logical flow. The semantics of video events that complement the features extracted by signal processing can be annotated using markup-based [16], ontology-based [17], and formal rule-based [18] representation.

The approaches to detect *regions of interest* (ROIs) fall into two categories. The *generalized approaches* are based on visual attention models that determine the likelihood of a human fixing his or her gaze on a particular position, such as the horizon, in a video sequence. Visual attention models are usually based on the features perceived by human vision, such as color, orientation, movement direction, and disparity, which can be combined into a saliency map to indicate the probability of pixel drawing attention [19]. Feature extraction can be extended with motion detection and face detection to obtain more advanced visual attention models [20]. Since the positions at which the persons are looking are determined by the task [21], detection accuracy can be improved if the task is considered during detection. This is one of the main motivations behind the *application-based approaches*, which predict the region of interest a priori for a particular application (e.g., human faces in a video conferencing application).

Human action plays an important part in complex video event understanding, where an action is a sequence of movements generated by a person during the performance of a task. Action recognition approaches differ in terms of spatial and temporal decomposition, action segmentation and recognition from continuous video streams, and handling variations of camera viewpoint. Conventional 2D video scenes often do not convey enough information for *human action recognition* (HAR). When motion is perpendicular to the camera plane, the 3D structure of the scene is needed, which is typically obtained using depth sensors. The information obtained from depth sensors is suitable for human body model representation and skeleton tracking using masked joint trajectories through action template learning [22]. Human action recognition can also be performed using depth motion maps (DMMs), which are formed by projecting the depth frames of depth video sequences onto three orthogonal Cartesian planes and considering the difference between two consecutive projected maps under each projection view throughout the depth video sequence [23]. Real-time human action recognition is then utilized by a collaborative representation classifier with a distance-weighted Tikhonov matrix.

Beyond RDB-D depth cameras and Kinect depth sensors, inertial sensors are also applied in human action recognition [24]. The information fusion of the input obtained from RGB video cameras, depth cameras, and inertial sensors can exploit the benefits of the different representations of 3D action data [25].

In contrast to the static background of news videos, real-world actions often occur in crowded environments, where the motion of multiple objects distracts the segmentation of a particular action. One way to handle crowd flows is to consider them as collections of local spatiotemporal motion patterns in the scene, whose variation in space and time can be modeled with a set of statistical models [26].

## 1.6  Conceptualization of Multimedia Contents

While images and audio files might contain embedded machine-readable metadata, such as the geo-coordinates in JPEG image files [27] or the performer in MP3 audio files, video files seldom have anything beyond generic technical metadata, such as resolution and length, that do not convey information about the actual content. Because of the lack of content descriptors, the meaning of video scenes cannot be interpreted by automated software agents. Search engines still rely heavily on textual descriptors, tags, and labels for multimedia indexing and retrieval, because text-based data is still the most robust content form, which can be automatically processed using natural language processing.

The application of well-established image processing algorithms is limited in automated video processing, because videos are far more complex than images. Although there are common challenges in image and video processing, such as occlusion, background clutter, pose and lighting changes, and in-class variation, videos have unique characteristics. In contrast to images, videos convey spatiotemporal information, are inherently ambiguous, usually have an audio channel, are often huge in size, and are open to multiple interpretations. Moreover, the compression algorithm used in a video file determines which frames can be used, making a careful selection necessary to prevent processing frames depicting a transition (e.g., face washed out from motion blur). On top of these challenges, video surveillance applications and robotic vision require real-time processing, which is challenging due to the computing complexity and enormous amount of data involved.

One of the approaches to address some of the aforementioned issues is to use *Semantic Web standards* to create *ontologies*, which provide a formal conceptualization of the intended semantics of a knowledge domain or common sense human knowledge, i.e., an abstract, simplified view of the world represented for a particular purpose.

The logical formalism behind web ontologies provides robust modeling capabilities with formal model-theoretic semantics. These semantics are defined as a model theory representing an analogue or a part of the world being modeled, where objects of the world are modeled as elements of a set, and the relationships between the objects as sets of tuples. To accommodate the various needs of applications, different sets of mathematical constructors can be implemented for concrete usage scenarios while establishing the desired level of expressivity, manageability, and computational complexity. The formal foundation of ontologies provides precise definition of relationships between logical statements, which describes the intended behavior of ontology-based systems in a machine-readable form. The logical underpinning of web ontologies is useful not only for the formal definition of concepts, but also for maintaining ontology consistency and integrating multiple ontologies. This data description formalism does not make the *unique name assumption* (UNA), i.e., two concepts with different names might be considered equivalent in some inferred statements. In addition, any true statement is also known to be true, i.e., if a fact is not known, the negation of the fact cannot be implied automatically (*closed world assumption*, CWA).

Web ontologies hold sets of machine-interpretable statements, upon which logical consequences can be inferred automatically (as opposed to modeling languages such as the Unified Modeling Language (UML)), which enables complex high-level scene interpretation tasks, such as recognizing situations and temporal events rather than just objects [28]. The most common *inference*[1] tasks have been implemented in the form of efficient decision procedures that, depending on the

---

[1]Deriving logical conclusions from premises known or assumed to be true.

complexity of the formalism, can even guarantee that they will return a Boolean true or false value for the decision problem in a predetermined timeframe, i.e., they will not loop indefinitely.

Ontology engineers have covered many different knowledge domains for multimedia content, thereby enabling the efficient representation, indexing, retrieval, and processing of, among others, medical videos, surveillance videos, soccer videos, and tennis videos. Nevertheless, the semantic enrichment of multimedia resources, which provides efficient querying potential, often requires human cognition and knowledge, making automated annotation inaccurate or infeasible. Considering the millions of multimedia files available online, manual annotation is basically infeasible, even though several social media portals support *collaborative annotation*, through which annotations of multimedia resources are dynamically created and curated by multiple individuals [29]. Manual annotation has many drawbacks, clearly indicated by the misspelt, opinion-based, vague, polysemous or synonymous, and often inappropriately labeled categories on video sharing portals such as YouTube.

By implementing Semantic Web standards according to best practices, the depicted concepts, properties, and relationships can be described with high-level semantics, individually identified on the Internet and interlinked with millions of related concepts and resources in a machine-interpretable manner. To minimize the long web addresses in knowledge representation, the namespace mechanism is frequently used to abbreviate domains and directories, and reveal the meaning of tags and attributes by pointing to an external vocabulary that describes the concepts of the corresponding knowledge domain in a machine-processable format. For example, a movie ontology defines movie features and the relationship between these features in a machine-processable format, so that software agents can "understand" their meanings (e.g., title, director, running time) in a dataset or on a web page by pointing to the ontology file.

Interlinking the depicted concepts with related concept definitions puts the depicted concepts into context, improves concept detection accuracy, eliminates ambiguity, and refines semantic relationships. Among other benefits, organized data support a wide range of tasks and can be processed very efficiently.

## 1.7   Concept Mapping

Computational models are used to map the multimedia features to concept definitions, which can eventually be exploited by hypervideo applications, search engines, and intelligent applications. Semantic concept detection relies on multimedia data for training typically obtained though low-level feature extraction and feature-based model learning. The efficiency of concept detection is largely determined by the availability of multimedia training samples for the knowledge domain most relevant to the depicted concepts. Because of the time-consuming nature of manual annotation, the number of labeled samples is often insufficient, which is partly addressed by semi-supervised learning algorithms, such as co-training [30].

While learning multimedia semantics can be formulated as supervised machine learning, not every machine learning algorithm is suitable due to the limited number of positive examples for each concept, incoherent negative examples, and the large share of overly generic examples. Oversampling can address some of these issues by replicating positive training examples, albeit it increases training data size and the time requirement of training. Another approach, undersampling, ignores some of the negative examples; however, this might result in losing some useful examples. To combine the benefits of the two approaches while minimizing their drawbacks, negative data can be first partitioned, then classifiers created based on positive examples and negative example groups [31].

The definition of semantics for the concepts depicted in a region of interest, keyframe, shot, video clip, or video, along with their properties and relationships, is provided by vocabularies and ontologies. Well-established common sense knowledge bases and ontologies that can be used for describing the concepts depicted in videos include *WordNet*[2] and *OpenCyc*.[3] General-purpose upper ontologies, such as *DOLCE*[4] and *SUMO*,[5] are also used in multimedia descriptions. Depending on the knowledge domain, correlations of concepts might be useful for improving the conceptualization of multimedia contents [32]. For example, a beach scene is far more likely to depict surfs, sand castles, and palm trees than a traffic scene, and so collections of concepts that often occur together provide additional information and context for scene interpretation. One of the most well-known ontologies to provide such predefined semantic relationships and co-occurrence patterns, although not without flaws, is the *Large-Scale Concept Ontology for Multimedia (LSCOM)* [33]. Class hierarchies of ontologies further improve scene interpretation. For example, the subclass-superclass relationships between concepts of an animal ontology make it machine-interpretable that a koala is a mammal, therefore both concepts are correct for the concept mapping of a depicted koala, only the first one is more specific than the second one. Moreover, multimedia concepts are usually not isolated, and multiple concepts can be associated with any given image or video clip, many of which are frequently correlated. For example, a koala is very likely to be depicted together with a eucalyptus tree, but more than unlikely with a space shuttle. In ontology-based scene interpretation, the a priori and asserted knowledge about a knowledge domain can be complemented by rule-based, inferred statements [34].

---

[2]http://wordnet-rdf.princeton.edu/ontology

[3]https://sourceforge.net/projects/texai/files/open-cyc-rdf/1.1/

[4]http://www.loa.istc.cnr.it/old/ontologies/DLP3971.zip

[5]http://www.adampease.org/OP/

## 1.8   Implementation Potential: From Search Engines to Hypervideo Applications

The machine-interpretable descriptions created using Semantic Web standards provide universal access to multimedia data for humans and computers alike. The unique identifiers used by these descriptions enable the separation of the description from the multimedia content, which is very beneficial in multimedia retrieval, because small text files are significantly easier to transfer and process than the actual multimedia files and encourage data reuse instead of duplicating data. The formal concept definitions eliminate ambiguity in these descriptors, and their interlinking makes them very efficient in finding related concepts. Furthermore, huge multimedia files have to be downloaded only if they seem to be truly relevant or interesting to the user. These descriptors can be distributed in powerful purpose-built databases and embedded directly in the website markup as lightweight annotations to reach the widest audience possible, providing data for state-of-the-art search engine optimization.

Semantics enable advanced applications that exploit formal knowledge representation. Such computer software can provide fully customized interfaces to service subscribers, automatically identify suspicious activity in surveillance videos, classify Hollywood movies, generate age rating for movies, and identify previously unknown risk factors for diseases from medical videos.

The semantically enriched multimedia contents can be searched using multimedia search terms, somewhat similar to searching text files. For example, users can find music that actually sounds similar (have similar frequencies, wavelengths, instruments, etc.) to the music they like. Videos can be searched for certain clips or a particular kind of movement. Hypervideo applications can play videos while displaying information about their content, position the playback to a particular part of a video based on semantics, and so on.

## 1.9   Summary

This chapter listed the main challenges of machine interpretation of images and video scenes. It highlighted the limitations of low-level feature-based classification, object recognition, and multimedia understanding. The utilization of the conceptualization of multimedia contents in search engines for content-based multimedia retrieval and hypervideo applications was also discussed.

# Chapter 2
# Knowledge Representation with Semantic Web Standards

The content of conventional websites is human-readable only, which is not suitable for automated processing and inefficient when searching for related information. Traditional web datasets can be considered as isolated data silos that are not linked to each other. This limitation can be addressed by organizing and publishing data using powerful formats that add structure and meaning to the content of web pages and interlink related data. Computers can interpret such data better, which enables task automation. To improve the automated processability of multimedia content, formal knowledge representation standards are required. These standards can be used not only to annotate multimedia resources with machine-readable data, but also to express complex statements about them and their relationships.

## 2.1 The Semantic Web

The websites that provide semantics (meaning) to software agents form the Semantic Web, an extension of the conventional Web [35] introduced in the early 2000s [36]. To improve the machine processability of digital resources, formal knowledge representation standards are required, which enable the interpretation of concepts and their relationships for software agents. Machine-interpretable definitions are used on different levels of conceptualization and comprehensiveness, resulting in the following types of *knowledge organization systems (KOS)*:

- *Thesauri:* reference works that list words grouped together according to similarity of meaning, including synonyms and sometimes antonyms
- *Taxonomies:* categorized words with hierarchies that can be extracted automatically
- *Knowledge bases (KB):* terminology and individuals of a knowledge domain, including the relationships between them

© Springer International Publishing AG 2017
L.F. Sikos, *Description Logics in Multimedia Reasoning*,
DOI 10.1007/978-3-319-54066-5_2

- *Ontologies:* formal conceptualizations of a knowledge domain with complex relationships and complex rules suitable for inferring new statements automatically.[1] The term "ontology" was originally used in philosophy to study the nature of existence. The first implementations in computer science appeared in the 1980s and considered ontology as "a conception of reality, a model of the world," such as in the form of a linguistic sensory-motor ontology, which can be used for natural language processing and utilized in robotics to provide intelligent output (e.g., movements of robotic arms) [37]. Ontologies with machine-readable statements have been introduced in artificial intelligence in the early 1990s to specify content-specific agreements for sharing and reuse of formally represented human knowledge among computer software [38].
- *Datasets:* collections of related machine-interpretable statements

## 2.2   Unstructured, Semistructured, and Structured Data

Text files, conventional websites, and textual descriptions of multimedia resources constitute *unstructured data*, which is human-readable only [39]. As a simplistic example, assume a short natural language description of a movie that includes the title and the running time, for example, "The Notebook is 124 minutes long and starring Ryan Gosling." Software agents, such as search engines, consider this string as meaningless consecutive characters. In fact, the word "notebook" is ambiguous with multiple meanings, and so without context a software agent might not be able to determine whether the above sentence refers to a small binder of paper pages to take notes or a portable computer. In a semistructured, well-formed XML document, which follows rigorous syntax rules, this example can be written as shown in Listing 2.1.

**Listing 2.1**  A well-formed XML example

```
<?xml version="1.0" encoding="UTF-8"?>
<movie xmlns:xsd="http://www.w3.org/2001/XMLSchema#">
   <title type="xsd:string">The Notebook</title>
   <duration type="xsd:positiveInteger">124</duration>
   <starring type="xsd:string">Ryan Gosling</starring>
</movie>
```

This code is machine-readable, and so computers can extract different entities and properties from the text, such as 124 is understood as a positive integer; however, it still does not have a meaning. To overcome this limitation, contents

---

[1]It can be confusing, but a knowledge organization formalism on the Semantic Web might be called a knowledge base in the representation language and an ontology in the implementation language, thereby fading the distinction between the two terms (see later in Chap. 4).

can be made machine-processable and unambiguous by leveraging organized, *structured data*. Structured data enables a much wider range of automated tasks than unstructured data, and structured data can be processed far more efficiently than plain text files or relational databases. The previous code can be made machine-interpretable by extending it with annotations from a knowledge organization system, such as a *controlled vocabulary* (see Definition 2.6), which collects terms (*concepts*, see Definition 2.2), properties and relationships (*roles*, see Definition 2.3), and objects (*individuals*, see Definition 2.1) of a *knowledge domain* (field of interest or area of concern) using *Internationalized Resource Identifiers* (IRIs, see Definition 2.5) or any valid subsets of IRIs, such as *Uniform Resource Identifiers* (URIs, see Definition 2.4).

**Definition 2.1 (Individual)**   An individual is a constant representing a real-world entity.

**Definition 2.2 (Concept)**   A concept is a unary predicate representing a class of real-world entities, which is either an atomic concept (concept name) $A$ denoting a set of individuals, or a nonatomic (complex) concept $C$ expressed as a compound statement, such as conjunction, disjunction, and negation of atomic concepts of the form $A \sqcap B$, $A \sqcup B$, and $\neg A$, and existential and value restrictions of the form $\exists R.C$ and $\forall R.C$, where $A$ and $B$ represent atomic concepts, $C$ represents a (possibly) complex concept, and $R$ represents a role. The list of available complex concepts depends on the knowledge representation language.

**Definition 2.3 (Role)**   A role is a binary predicate $R$ that captures a relationship between, or defines a property of, concepts and individuals, and optionally defines the source and target concepts or individuals, and, when applicable, the datatype (or datatype restriction) for its permissible values.

**Definition 2.4 (Uniform Resource Identifier, URI)**   A URI is a string of ASCII characters used to identify a resource in the form `scheme:[//[user:pass-word@]host[:port]][/]path[?query][#fragment]`. The set of URIs is typically denoted as $\mathbb{U}$.

**Definition 2.5 (Internationalized Resource Identifier, IRI)**   An IRI is a string of Unicode characters used to identify a resource in the form `scheme:[//[user:password@]host[:port]][/]path[?query][#fragment]`. The set of IRIs is typically denoted as $\mathbb{I}$.

**Definition 2.6 (Controlled Vocabulary)**   A controlled vocabulary is a triple $V = (N_C, N_R, N_I)$ of countably infinite sets of IRI symbols denoting atomic concepts (concept names or classes) ($N_C$), atomic roles (role names, properties, or predicates)

$(N_R)$, and individual names (objects) $(N_I)$, respectively, where $N_C$, $N_R$, and $N_I$ are pairwise disjoint sets.

Compared to controlled vocabularies, ontologies leverage far more constructors. They declare not only atomic concepts, atomic roles, and individuals, but also reflexive, symmetric, irreflexive roles, disjointness, universal and existential quantification, and complex rules.[2]

The terms of controlled vocabularies and ontologies can be used to write machine-interpretable statements about any kind of resource. For example, to describe a video clip in a machine-interpretable format, a vocabulary or ontology that has the formal definition of movies is needed, such as Schema.org, which has a `Clip` class[3] that defines typical properties of movie clips, including, but not limited to, the actors starred in the video clip, the director, the genre, or the composer of the soundtrack of the clip.

## 2.3  RDF

One of the most fundamental standards of the Semantic Web is the *Resource Description Framework (RDF)*, a powerful knowledge representation language. The structured data of ontologies, knowledge bases, and datasets, as well as the lightweight annotations of websites are usually expressed in, or based on, one of the syntax notations or data serialization formats of RDF. RDF is suitable for creating machine-interpretable descriptions about any digital resource by incorporating factual data and definitions from arbitrary external vocabularies, knowledge bases, ontologies, and datasets. Which vocabulary or ontology to use depends on the area of interest to represent; however, some knowledge domains such as persons and movies can be described with classes and properties from more than one vocabulary. One can also create new ontologies for those knowledge domains that are not covered by existing ontologies.

The RDF standard provides a vocabulary and constructors for machine-interpretable statements, a graph representation for statements with IRIs, plain and typed literals, blank nodes (for anonymous resources), and several syntaxes.

The *RDF vocabulary (rdfV)* is a set of IRI references in the RDF namespace, `http://www.w3.org/1999/02/22-rdf-syntax-ns#`. The `rdf:type` predicate expresses class membership; `rdf:Property` is the class of properties; `rdf:XMLLiteral` is the datatype of XML literals; and `rdf:List`, `rdf:first`, `rdf:rest`, and `rdf:nil` can be used to write lists (*RDF collections*). Ordered, unordered, and alternate containers of values can be described using `rdf:Seq`,

---

[2]The formal definition of an ontology depends on the logical underpinning. Specific definitions are provided in Chap. 4.

[3]http://schema.org/Clip

rdf:Bag, rdf:Alt, rdf:_1, rdf:_2. *Reification*[4] can be expressed with rdf:Statement, rdf:subject, rdf:predicate, and rdf:object. Structured values are described using rdf:value.

The RDF data model is based on statements to describe digital resources in the form of subject–predicate–object (resource–property–value) expressions called *RDF triples* or *RDF statements* (see Definition 2.9), which might contain *RDF literals* (see Definition 2.7) and *blank nodes* (see Definition 2.8), such as Movie-Clip–runningTime–124 and video.mp4–title–"The Notebook."

**Definition 2.7 (RDF Literal)** An RDF literal is either (1) a self-denoting *plain literal* of the form "<string>"(@<lang>)?, where <string> is a string and <lang> is an optional language tag, or (2) a *typed literal* of the form "<string>"^^<datatype>, where <datatype> is an IRI denoting a datatype according to a schema, such as the XML Schema, and <string> is an element of the lexical space corresponding to the datatype. A typed literal denotes the value derived from the lexical to value mapping of <datatype> to <string>. The typical annotation of the set of all literals is $\mathbb{L}$, which consists of the set $\mathbb{L}_P$ of plain literals and the set $\mathbb{L}_T$ of typed literals, respectively.

Not every resource is an IRI or an RDF literal, because RDF statements can be written with unknown subjects or unknown objects as well, although this has to be avoided whenever possible.

**Definition 2.8 (Blank Node)** A blank node (bnode) is a unique resource, which is neither an IRI nor an RDF literal. The set of all blank nodes is usually denoted as $\mathbb{B}$.

Blank nodes can only be identified by their properties or relations, and cannot be named directly. For example, if one knows that his distant relative "Tom" has a child, but the name of the child is unknown, the child can be represented as a blank node. When describing a family tree, such a formalism would still indicate that Tom is a father if the hasChild relation is used to identify parents.

Using the definition of IRIs, RDF literals, and blank nodes, RDF triples can be formally defined (see Definition 2.9).

**Definition 2.9 (RDF Triple)** Assume there are pairwise disjoint infinite sets $\mathbb{I}$, $\mathbb{L}$, and $\mathbb{B}$ representing IRIs, RDF literals, and blank nodes, respectively. A triple $(s, o, p) \in (\mathbb{I} \cup \mathbb{B}) \times \mathbb{I} \times (\mathbb{I} \cup \mathbb{L} \cup \mathbb{B})$ is called an RDF triple (or RDF statement), where $s$ is the subject, $p$ the predicate, and $o$ the object.

---

[4]Reification is making an RDF statement about another statement. It is useful for expressing the time when a statement was made, the person who made the statement and how certain she is about it, how relevant the relation is, and so on. However, reification has some known issues, as it can cause redundancy and the proliferation of objects, and can lead to limited reasoning potential or even contradiction.

The predicate annotates a relationship between the subject and the object, which addresses a limitation of HTML and XML. Listing 2.1 provides hierarchy; however, it is out of context. The external vocabulary that defines movies and movie properties has to be declared using the namespace mechanism. In RDF/XML serialization, for example, this can be done using the xmlns attribute (see Listing 2.2).

**Listing 2.2** RDF/XML serialization example

```
<?xml version="1.0" encoding="UTF-8" ?>
<rdf:RDF
  xmlns:dc="http://purl.org/dc/elements/1.1/"
  xmlns:rdf="http://www.w3.org/1999/02/22-rdf-syntax-ns#"
  xmlns:schema="http://schema.org/"
  xmlns:vidont="http://vidont.org/"
  xmlns:xsd="http://www.w3.org/2001/XMLSchema#">
  <rdf:Description rdf:about="http://example.com/video.mp4">
    <rdf:type>schema:MovieClip</rdf:type>
    <dc:title>The Notebook</dc:title>
    <vidont:runningTime>124</vidont:runningTime>
    <vidont:starring>Ryan Gosling</vidont:starring>
  </rdf:Description>
</rdf:RDF>
```

To shorten long and/or frequently used URIs, the RDF statements can be abbreviated using the namespace mechanism, which concatenates the prefix URI and the vocabulary or ontology term. In this case, the schema: prefix is used to abbreviate http://schema.org/, and so schema:MovieClip stands for http://schema.org/MovieClip. Similarly, dc:title abbreviates http://purl.org/dc/terms/title. Listing 2.2 describes an online video resource (video.mp4, the subject) as a movie clip (http://schema.org/MovieClip, the object) using the very common rdf:type predicate (here expressing an "is a" relationship, i.e., video.mp4 "is a" movie clip). It also declares the movie title as The Notebook using the definition of title from Dublin Core, and the running time and an actor who starred in the movie using their definition from the *Video Ontology (VidOnt)*[5] (http://vidont.org/runningTime and http://vidont.org/starring, respectively).[6]

---

[5]http://vidont.org/vidont.ttl

[6]These links are often symbolic links that do not always point to a dedicated web page for each individual property, and are sometimes forwarded to the domain of the namespace. Some vocabularies have a namespace address mechanism whereby all links point directly to the corresponding section of the machine-readable vocabulary file. The human-readable explanation of the properties of external vocabularies is not always provided. The namespace prefix ends in either a slash (/) or a hash (#). Note that these links are usually case-sensitive.

The format and serialization of the structured data are independent of the vocabulary definitions, therefore an RDF representation can be written not only in the RDF/XML serialization used in Listing 2.2, but also in other RDF serializations. As an example, Listing 2.3 shows the Turtle equivalent of Listing 2.2.

**Listing 2.3** Turtle serialization example

```
@prefix dc: <http://purl.org/dc/elements/1.1/> .
@prefix rdf: <http://www.w3.org/1999/02/22-rdf-syntax-ns#> .
@prefix schema: <http://schema.org/> .
@prefix vidont: <http://vidont.org/> .
@prefix xsd: <http://www.w3.org/2001/XMLSchema#> .

<http://example.com/video.mp4> rdf:type schema:MovieClip ;
dc:title "The Notebook" ; vidont:runningTime
"124"^^xsd:integer ; vidont:starring "Ryan Gosling" .
```

As you can see, this syntax is easier to read and more compact than RDF/XML.[7] In fact, some parts of the above code can be abbreviated even further, such as `rdf:type` can simply be written as a and `"124"^^xsd:integer` as `124`, as per Turtle specifications.[8] Further syntax notations and data serialization formats for RDF include RDFa,[9] JSON-LD,[10] HTML5 Microdata,[11] Notation3 (N3),[12] N-Triples,[13] TRiG,[14] and TRiX.[15] Regardless of the serialization used, all RDF triples can be visually represented as an *RDF graph* (see Definition 2.10).

**Definition 2.10 (RDF Graph)** An RDF graph is a set of RDF triples, where the set of nodes is the set of subjects and objects of RDF triples in the graph.

The RDF graphs that do not have blank nodes are called *ground RDF graphs* (see Definition 2.11).

**Definition 2.11 (Ground RDF Graph)** A ground RDF graph over $\mathbb{I}, \mathbb{L}, \mathbb{B}$ is a set of ground triples over $\mathbb{I}, \mathbb{L}, \mathbb{B}$, which are elements in $(\mathbb{I} \times \mathbb{I} \times (\mathbb{I} \cup \mathbb{L}))$.

Figure 2.1 visualizes the previous example as an RDF graph.

---

[7]Because of its brevity, simplicity, and popularity, the Turtle syntax is used throughout this book for all the RDF and RDF-based examples, except where other serializations are demonstrated.

[8]https://www.w3.org/TR/turtle/

[9]https://www.w3.org/TR/rdfa-primer/

[10]https://www.w3.org/TR/json-ld/

[11]https://www.w3.org/TR/microdata/

[12]https://www.w3.org/TeamSubmission/n3/

[13]https://www.w3.org/TR/n-triples/

[14]https://www.w3.org/TR/trig/

[15]http://www.hpl.hp.com/techreports/2003/HPL-2003-268.pdf

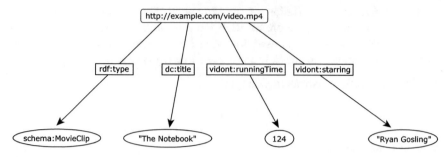

**Fig. 2.1**  An RDF graph

The semantics of RDF graphs is fixed via interpretations, such as *simple RDF interpretations* (see Definition 2.12) and *RDF interpretations*.

**Definition 2.12 (Simple RDF Interpretation)**  A simple interpretation of an RDF vocabulary of IRIs and literals, $V = \mathbb{I} \cup \mathbb{L}$, is a sextuple (*IR, IP, IEXT, IS, IL, LV*), where $IR = \varnothing$ is the domain of interpretation $I$, $IP$ the set of properties of interpretation $I$, $IEXT : IP \rightarrow P(IR \times IR)$, $IS : \mathbb{I} \rightarrow IR \cup IP$, $IL$ a mapping from typed literals in $V$ into $IR$, and $LV \subseteq IR$ the set of literal values.

Simple RDF interpretations only provide the semantics of URIs/IRIs and typed literals. The semantics of RDF triples, graphs, and untyped literals are provided by *RDF denotations*, which define the truth values for triples and graphs by recursively applying interpretations, as shown in Definition 2.13.

**Definition 2.13 (Denotation of Ground RDF Graphs)**  Given an RDF interpretation $I = (IR, IP, IEXT, IS, IL, LV)$ over a vocabulary $V$, the denotation of a ground RDF graph is defined as follows:

–  If $E$ is a plain literal "*aaa*" in $V$, then $I(E) = \texttt{aaa}$
–  If $E$ is a plain literal "*aaa*"@*ttt* in $V$, then $I(E) = \texttt{<aaa,ttt>}$
–  If $E$ is a typed literal in $V$, then $I(E) = IL(E)$
–  If $E$ is an IRI reference in $V$, then $I(E) = IS(E)$
–  If $E$ is a ground triple of the form $(s,p,o)$, then $I(E)$ is true if $s, p, o$ are in $V, I(p)$ is in $IP$, and $(I(s),I(o))$ is in $IEXT(I(p))$
–  If $E$ is a ground RDF graph, then $I(E)$ is false if $I(E')$ is false for some triple $E'$ in $E$, otherwise $I(E)$ is true

*RDF interpretations* extend simple RDF interpretations by further restricting the usage of the `type` predicate and XML literals [40].

## 2.4 RDFS

While RDF is the cornerstone of the Semantic Web, it was designed to describe machine-interpretable statements, and not to formally define terms of a knowledge domain. For this reason, the RDF vocabulary was extended by concepts required for creating controlled vocabularies and basic ontologies, resulting in the *RDF Schema Language (RDFS)*, which was later renamed to the *RDF Vocabulary Description Language*. RDFS is suitable for defining terms of a knowledge domain and basic relationships between them [41]. RDFS-based vocabularies and ontologies can be represented as RDF graphs.

The RDFS classes and properties form the *RDFS vocabulary (rdfsV)*, but being an extension of RDF, RDFS also reuses properties from the RDF vocabulary. The namespace URI of RDFS is `http://www.w3.org/2000/01/rdfschema#`, which is abbreviated with the `rdfs:` prefix. The RDFS vocabulary defines class resources (`rdfs:Resource`), the class of literal values such as strings and integers (`rdfs:Literal`), the class of classes (`rdfs:Class`), the class of RDF datatypes (`rdfs:Datatype`), the class of RDF containers (`rdfs:Container`), and the class of container membership properties (`rdfs:ContainerMembershipProperty`). The properties of RDFS can express that the subject is a subclass of a class (`rdfs:subClassOf`), the subject is a subproperty of a property (`rdfs:subPropertyOf`), add a human-readable name for the subject (`rdfs:label`), declare a description of the subject resource (`rdfs:comment`), identify a member of the subject resource (`rdfs:member`), add information related to the subject resource (`rdfs:seeAlso`), and provide the definition of the subject resource (`rdfs:isDefinedBy`).

Properties can be declared to apply to only certain instances of classes by defining their domain and range that indicate the relationships between RDFS classes and properties and RDF data. The `rdfs:domain` predicate indicates that a particular property applies to instances of a designated class (the domain of the property), in other words declares the class of those resources that may appear as subjects in a triple with the predicate. The `rdfs:range` predicate indicates that the values of a particular property are instances of a designated class or its allowed values are of a certain datatype, i.e., the class or datatype of those resources that may appear as the object in a triple with the predicate, also known as the range of the property, as shown in Listing 2.4.

**Listing 2.4** Using RDFS domain and range

```
@prefix rdf: <http://www.w3.org/1999/02/22-rdf-syntax-ns#> .
@prefix rdfs: <http://www.w3.org/2000/01/rdf-schema#> .
@prefix schema: <http://schema.org/> .
@prefix vidont: <http://vidont.org/> .

schema:Movie a rdfs:Class .
schema:Person a rdfs:Class .
```

```
vidont:Unforgiven a schema:Movie .
vidont:ClintEastwood a schema:Person .
vidont:directedBy rdfs:domain schema:Movie .
vidont:directedBy rdfs:range schema:Person .
vidont:Unforgiven vidont:directedBy vidont:ClintEastwood .
```

The *RDFS interpretation* is based on the RDF interpretation (see Definition 2.14).

**Definition 2.14 (RDFS Interpretation)** An RDFS interpretation is an RDF interpretation that satisfies the following additional conditions and the RDFS axiomatic triples:

- $x \in ICEXT(y)$ iff $(x,y) \in IEXT(I(type))$
- $IC = ICEXT(I(Class))$
- $IR = ICEXT(I(Resource))$
- $LV = ICEXT(I(Literal))$
- If $(x,y)$ is in $IEXT(I(domain))$ and $(u,v)$ is in $IEXT(x)$, then $u$ is in $ICEXT(y)$
- If $(x,y)$ is in $IEXT(I(range))$ and $(u,v)$ is in $IEXT(x)$, then $v$ is in $ICEXT(y)$
- $IEXT(I(subPropertyOf))$ is transitive and reflexive on $IP$
- If $(x,y)$ is in $IEXT(I(subPropertyOf))$, then $x$ and $y$ are in $IP$ and $IEXT(x)$ is a subset of $IEXT(y)$
- If $x$ is in $IC$, then $(x,I(Resource))$ is in $IEXT(I(subClassOf))$
- If $(x,y)$ is in $IEXT(I(subClassOf))$, then $x$ and $y$ are in $IC$ and $ICEXT(x)$ is a subset of $ICEXT(y)$
- $IEXT(I(subClassOf))$ is transitive and reflexive on $IC$
- If $x$ is in $ICEXT(I(ContainerMembershipProperty))$, then $(x,I(member))$ is in $IEXT(I(subPropertyOf))$
- If $x$ is in $ICEXT(I(Datatype))$, then $(x, I(Literal))$ is in $IEXR(I(subClassOf))$

## 2.5  OWL

While simple machine-readable ontologies can be created using RDFS, complex knowledge domains require more capabilities, such as the following:

- Relations between classes (union, intersection, disjointness, equivalence)
- Property cardinality constraints (minimum, maximum, exact number, e.g., a person has exactly one father)
- Rich typing of properties
- Characteristics of properties and special properties (transitive, symmetric, functional, inverse functional, e.g., A ex:hasAncestor B and B ex:hasAncestor C implies that A ex:hasAncestor C)
- Specifying that a given property is a unique key for instances of a particular class
- Domain and range restrictions for properties when they are used with a certain class

- Equality of classes, specifying that two classes with different URI references actually represent the same class
- Equality of individuals, specifying that two instances with different URI references actually represent the same individual
- Enumerated classes

The *Web Ontology Language* (intentionally abbreviated as *OWL* rather than WOL [42]) is a knowledge representation language that semantically extends RDF, RDFS, and its predecessor language, DAML+OIL. OWL was specially designed for creating web ontologies with a rich set of modeling constructors, addressing the ontology engineering limitations of RDFS. The development of the first version of OWL was started in 2002, and the second version, *OWL 2*, in 2008. OWL became a W3C Recommendation in 2004 [43], and OWL 2 was standardized in 2009 [44]. OWL 2 is a superset of OWL, and so all OWL ontologies can be handled by both OWL and OWL 2 applications. The significant extensions of OWL 2 over OWL include new constructors that improve expressiveness, extended datatype support, extended annotation capabilities, *syntactic sugar* (shorthand notation for common statements), and three profiles (fragments or sublanguages). Each OWL 2 profile provides a different balance between expressive power and reasoning complexity, thereby providing more options for different implementation scenarios [45]:

- *OWL 2 EL*: designed for handling ontologies with very large numbers of properties and/or classes
- *OWL 2 QL*: aimed at applications with a very large instance data volume and a priority for query answering
- *OWL 2 RL*: designed for applications that require scalable reasoning with relatively high expressivity

The default namespace of OWL and OWL 2 is `http://www.w3.org/2002/07/owl#`, which defines the OWL vocabulary. There is no MIME type dedicated to OWL, but the `application/rdf+xml` or the `application/xml` MIME type is recommended for XML-serialized OWL documents with the `.rdf` or `.owl` file extension. Turtle files use the `text/turtle` MIME type and the `.ttl` file extension.

### 2.5.1  OWL Variants

There are three flavors of OWL, each constituting different compromises between expressivity and computational complexity:

- *OWL Full* has no restrictions on the use of language constructors. It provides maximum expressiveness, syntactic freedom, but without computational guarantees. The semantics of OWL Full is a mixture of RDFS and OWL DL (RDF-based semantics).

- *OWL DL* is a restricted version of OWL Full. OWL DL provides very high expressiveness, computational completeness (all conclusions are guaranteed to be computable), and decidability (all computations can be finished in finite time). While OWL DL includes all OWL language constructors, they can be used only under certain restrictions. For example, OWL DL number restrictions may not be assigned to transitive properties.
- *OWL Lite* is a subset of OWL DL designed for easy implementation. OWL Lite has limited applicability, because it is suitable only for classification hierarchies and simple constraints.

All three flavors are available in both OWL and OWL 2.

### 2.5.2   Modeling with OWL

OWL property declarations use `rdf:type` as the predicate and the property type as the object. In OWL, there are four different property types:

- *Ontology properties* (`owl:OntologyProperty`) include the ontology import property (`owl:imports`) and ontology versioning properties (`owl:priorVersion`, `owl:backwardCompatibleWith`, and `owl:incompatibleWith`).
- *Annotation properties* declare annotations, such as labels, comments, and related resources (`owl:AnnotationProperty`). Five annotation properties are predefined in OWL, namely, `owl:versionInfo`, `rdfs:label`, `rdfs:comment`, `rdfs:seeAlso`, and `rdfs:isDefinedBy`.
- *Object properties* link individuals to other individuals (`owl:ObjectProperty`). The object property highest in the hierarchy is `owl:topObjectProperty`.
- *Datatype properties* link individuals to data values (`owl:DatatypeProperty`).

The *ontology header*, which provides general information about an ontology, employs ontology properties and annotation properties. OWL ontologies are identified globally in ontology files by declaring the URI as the subject, `rdf:type` as the predicate, and `owl:Ontology` as the object (see Listing 2.5).

**Listing 2.5** Ontology declaration

```
@prefix rdf: <http://www.w3.org/1999/02/22-rdf-syntax-ns#> .
@prefix owl: <http://www.w3.org/2002/07/owl#> .

<http://example.com/flowerontology.owl> a owl:Ontology .
```

Using this resource identifier as the subject, general information defined in other ontologies can be reused in the ontology by using `owl:imports` as the predicate, and the resource identifier of the other ontology as the object (see Listing 2.6).

**Listing 2.6** Importing an ontology

```
@prefix owl: <http://www.w3.org/2002/07/owl#> .

<http://example.com/flowerontology.owl> owl:imports
<http://example.org/plantontology.owl> .
```

Further information about an ontology can be defined in the ontology header using the owl:versionInfo, rdfs:label, rdfs:comment, rdfs:seeAlso, and rdfs:isDefinedBy annotation properties.

The following sections demonstrate the most common class, property, and individual declaration types available in OWL 2.

### 2.5.2.1 Class Declaration

Classes are declared in OWL by using the rdf:type predicate and setting the object to owl:Class. Class hierarchy is declared in the form of subclass-superclass relationships using the rdfs:subClassOf predicate. Atomic classes can be combined into complex classes using owl:equivalentClass, which contains all instances of both classes (owl:intersectionOf), every individual that is contained in at least one of the classes (owl:unionOf), and instances that are not instances of a class (owl:complementOf) (see Listing 2.7).

**Listing 2.7** Atomic class, complex class, and subclass declaration in OWL

```
@prefix owl: <http://www.w3.org/2002/07/owl#> .
@prefix rdf: <http://www.w3.org/1999/02/22-rdf-syntax-ns#> .
@prefix rdfs: <http://www.w3.org/2000/01/rdf-schema#> .
@prefix schema: <http://schema.org/> .
@prefix vidont: <http://vidont.org/> .

vidont:Video a owl:Class .
vidont:Frame a owl:Class ; rdfs:subClassOf vidont:Video .
vidont:Keyframe a owl:Class ; rdfs:subClassOf vidont:Frame .
schema:BroadcastChannel a owl:Class ; owl:equivalentClass [ a
owl:Class ; owl:unionOf ( vidont:webcastChannel
schema:TelevisionChannel ) ] .
```

Not every individual can be an instance of multiple classes: some class memberships exclude membership in other classes, i.e., instances cannot belong to two disjoint classes. For example, music videos cannot be movie trailers (see Listing 2.8).

**Listing 2.8** Declaration of disjoint classes

```
@prefix owl: <http://www.w3.org/2002/07/owl#> .
@prefix rdf: <http://www.w3.org/1999/02/22-rdf-syntax-ns#> .
@prefix schema: <http://schema.org/> .

[] a owl:AllDisjointClasses ;
owl:members ( schema:MusicVideoObject schema:trailer ) .
```

#### 2.5.2.2　Property Declaration

Ontology axioms employ object properties to make statements about the relationship between properties. Datatype properties define the datatype of a property, enumerate the permissible values of a property, or define an interval of permissible values for a property. Object and datatype properties are also used for defining a special subclass of a class based on particular property values and value ranges. Annotation properties are used to declare labels, comments, and ontology versioning information. In OWL DL, annotation properties are not allowed in axioms, but only in the ontology header.[16]

Property hierarchy is declared in the form of subproperty-superproperty relationships using `rdfs:subPropertyOf`. For example, in the VidOnt video ontology, the `isAttackedBy` property is defined as a subproperty of the `isActing` object property (see Listing 2.9).

**Listing 2.9** Subproperty declarations

```
@prefix owl: <http://www.w3.org/2002/07/owl#> .
@prefix rdfs: <http://www.w3.org/2000/01/rdf-schema#> .
@prefix vidont: <http://vidont.org/> .

vidont:isActing a owl:ObjectProperty ; rdfs:subPropertyOf
owl:topObjectProperty .
vidont:isAttackedBy a owl:ObjectProperty ; rdfs:subPropertyOf
vidont:isActing.
```

The `videoBitrate` datatype property must have positive values and is defined as a subproperty of `hasVideoChannel`, which is a Boolean datatype property (see Listing 2.10).

---

[16]In OWL Full, there is no such restriction on the use of annotation properties.

**Listing 2.10** Property restrictions

```
@prefix rdfs: <http://www.w3.org/2000/01/rdf-schema#> .
@prefix owl: <http://www.w3.org/2002/07/owl#> .
@prefix vidont: <http://vidont.org/> .
@prefix xsd: <http://www.w3.org/2001/XMLSchema#> .

vidont:hasVideoChannel a owl:DataProperty ; rdfs:range
xsd:boolean .
vidont:videoBitrate a owl:DatatypeProperty ;
rdfs:subPropertyOf vidont:hasVideoChannel ; rdfs:range
xsd:positiveInteger .
```

Permissible values for datatype properties can be defined not only by external datatypes, such as the XSD datatypes used in the previous examples, but also by custom datatypes defined especially for the domain of an ontology by constraining or combining existing datatypes.

Restrictions on existing datatypes define an interval of values, which can be left-open (no minimum element defined), right-open (no maximum element defined), open (neither minimum nor maximum element defined), left-closed (minimum element defined), right-closed (maximum element defined), or closed (both minimum and maximum element defined). The restrictions are declared using `owl:withRestrictions`, and the restriction type is typically declared using a combination of `xsd:minInclusive`, `xsd:minExclusive`, `xsd:maxInclusive`, and `xsd:maxExclusive` (see Listing 2.11).

**Listing 2.11** Defining a new datatype by restricting permissible values of a de facto standard datatype

```
@prefix ex: <http://example.com/> .
@prefix rdf: <http://www.w3.org/1999/02/22-rdf-syntax-ns#> .
@prefix rdfs: <http://www.w3.org/2000/01/rdf-schema#> .
@prefix owl: <http://www.w3.org/2002/07/owl#> .
@prefix xsd: <http://www.w3.org/2001/XMLSchema#> .

ex:zeroone a owl:equivalentClass
  [ a rdfs:Datatype;
    owl:onDatatype xsd:float;
    owl:withRestrictions (
    [ xsd:minInclusive "0"^^xsd:float ]
    [ xsd:maxInclusive "1"^^xsd:float ]
    )
  ] .
```

Some properties can have only one unique value for each individual. Such properties are called *functional properties*. Both object properties and datatype properties can be functional properties. For example, to express that only one date can be associated with all the actors when they started acting, the `startedActing` datatype property in a movie ontology can be declared as a functional property (see Listing 2.12).

**Listing 2.12** Functional property example

```
@prefix ex: <http://example.com/> .
@prefix rdf: <http://www.w3.org/1999/02/22-rdf-syntax-ns#> .
@prefix owl: <http://www.w3.org/2002/07/owl#> .

ex:startedActing a owl:FunctionalProperty .
```

*Property equivalence* can be described using `owl:equivalentProperty`, which can be used, for example, in a camera ontology to express that f-stop and aperture are the same (see Listing 2.13).

**Listing 2.13** Property equivalence example

```
@prefix camera: <http://example.com/camera/> .
@prefix rdf: <http://www.w3.org/1999/02/22-rdf-syntax-ns#> .
@prefix owl: <http://www.w3.org/2002/07/owl#> .

camera:f-stop a owl:DatatypeProperty ; owl:equivalentProperty
camera:aperture .
```

*Existential quantification* can capture incomplete knowledge, e.g., there exists a director for every movie (see Listing 2.14).

**Listing 2.14** Existential quantification example

```
@prefix ex: <http://example.com/> .
@prefix rdf: <http://www.w3.org/1999/02/22-rdf-syntax-ns#> .
@prefix owl: <http://www.w3.org/2002/07/owl#> .

ex:Director owl:equivalentClass [ a owl:Restriction ;
owl:onProperty ex:directorOf ; owl:someValuesFrom:Movie ] .
```

*Universal quantification* is suitable for describing a class of individuals for which all related individuals must be instances of a particular class, e.g., all members of a band are musicians (see Listing 2.15).

**Listing 2.15**  Universal quantification example

```
@prefix ex: <http://example.com/> .
@prefix rdf: <http://www.w3.org/1999/02/22-rdf-syntax-ns#> .
@prefix owl: <http://www.w3.org/2002/07/owl#> .

ex:Band a owl:Class ; owl:equivalentClass [ a owl:Restriction ;
owl:onProperty ex:hasMember ; owl:allValuesFrom ex:Musician ] .
```

In contrast to existential quantification and universal quantification, *property cardinality restrictions* can specify not only a restriction, but also a minimum, maximum, or exact number for the restriction. A cardinality constraint is called an *unqualified cardinality restriction* if it constrains the number of values of a particular property irrespective of the value type. In contrast, *qualified cardinality restrictions* constrain the number of values of a particular type for a property. As an example, Listing 2.16 expresses that interpreters speak at least two languages.

**Listing 2.16**  A qualified cardinality restriction with a left-closed interval

```
@prefix ex: <http://example.com/> .
@prefix rdf: <http://www.w3.org/1999/02/22-rdf-syntax-ns#> .
@prefix rdfs: <http://www.w3.org/2000/01/rdf-schema#> .
@prefix owl: <http://www.w3.org/2002/07/owl#> .

ex:Interpreter rdfs:subClassOf [ a owl:Restriction;
owl:onProperty ex:speaks ; owl:minQualifiedCardinality 2 ;
owl:onClass ex:Language ] .
```

Listing 2.17 shows an example of a cardinality constraint for which the exact number is known, namely, that trikes are vehicles with exactly three wheels.

**Listing 2.17**  A qualified cardinality restriction with an exact number

```
@prefix ex: <http://example.com/> .
@prefix rdf: <http://www.w3.org/1999/02/22-rdf-syntax-ns#> .
@prefix owl: <http://www.w3.org/2002/07/owl#> .
@prefix xsd: <http://www.w3.org/2001/XMLSchema#> .

ex:Trike a [ a owl:Restriction ; owl:qualifiedCardinality
"3"^^xsd:nonNegativeInteger ; owl:onProperty ex:hasWheels ;
owl:onClass ex:Vehicle ] .
```

*Inverse properties* can be declared using `owl:inverseOf` as the predicate, as demonstrated in Listing 2.18.

**Listing 2.18** Inverse property declaration

```
@prefix rdf: <http://www.w3.org/1999/02/22-rdf-syntax-ns#> .
@prefix owl: <http://www.w3.org/2002/07/owl#> .
@prefix vidont: <http://vidont.org/> .

vidont:directorOf a owl:ObjectProperty ; owl:inverseOf
vidont:directedBy .
```

In case the direction of the property does not matter, i.e., the property and its inverse coincide, the property is called *symmetric* and can be declared using `owl:SymmetricProperty`, such as `:marriedTo a owl:SymmetricProperty` . If the direction of the property matters, the property is called *asymmetric* and can be declared using `owl:AsymmetricProperty`, as, for example, `:hasChild a owl:AsymmetricProperty` .

Two properties are called *disjoint* if there are no two individuals that are interlinked by both properties. For example, no one can be the parent and the spouse of the same person, i.e., `:hasParent owl:propertyDisjoint-With :hasSpouse` .

The properties that relate everything to themselves are called *reflexive properties* and can be declared using `owl:ReflexiveProperty`, e.g., `foaf:knows a owl:ReflexiveProperty` . The properties by which no individual can be related to itself are called *irreflexive properties*, which can express, for example, that nobody can be his own parent, i.e., `:parentOf a owl:IrreflexiveProperty` .

Transitivity can be expressed using `owl:TransitiveProperty`. For example, declaring `basedOn` as a *transitive property* (`:basedOn a owl:TransitiveProperty .`) means that a sequel of a movie based on a depiction is also based on the novel the screen adaptation was based on.

If a simple object property expression, which has no transitive subproperties, is not expressive enough to capture a property hierarchy, *property chains* can be used. For example, a property chain axiom can express that the child of a child is a grandchild (see Listing 2.19).

**Listing 2.19** A property chain axiom

```
@prefix family: <http://example.com/family/> .
@prefix owl: <http://www.w3.org/2002/07/owl#> .

family:grandchildOf owl:propertyChainAxiom ( family:childOf
family:childOf ) .
```

The individuals that are connected by a property to themselves can be described via *self-restriction* using `owl:hasSelf` (see Listing 2.20).

**Listing 2.20** Self-restriction

```
@prefix ex: <http://example.com/> .
@prefix rdf: <http://www.w3.org/1999/02/22-rdf-syntax-ns#> .
@prefix owl: <http://www.w3.org/2002/07/owl#> .
@prefix xsd: <http://www.w3.org/2001/XMLSchema#> .


ex:Narcissist owl:equivalentClass [ a
owl:Restriction ; owl:onProperty ex:likes ;
owl:hasSelf "true"^^xsd:boolean ] .
```

### 2.5.2.3   Individual Declaration

OWL individuals can be declared using `rdf:type` as the predicate and `owl:NamedIndividual` as the object in an RDF triple. For example, Disney's *Zorro* can be added to a movie ontology as shown in Listing 2.21.

**Listing 2.21** Individual declaration

```
@prefix movie: <http://example.com/movie/> .
@prefix rdf: <http://www.w3.org/1999/02/22-rdf-syntax-ns#> .
@prefix owl: <http://www.w3.org/2002/07/owl#> .


movie:DisneysZorro a owl:NamedIndividual .
```

To express that an individual is an instance of a class, a statement is made by simply adding the individual as the subject, `rdf:type` as the predicate, and the class as the object (see Listing 2.22).

**Listing 2.22** Declaring an individual as an instance of a class

```
@prefix movie: <http://example.com/movie/> .
@prefix rdf: <http://www.w3.org/1999/02/22-rdf-syntax-ns#> .
@prefix schema: <http://schema.org/> .


movie:DisneysZorro a schema:TVSeries .
```

Always the most specific class should be declared, because the individual will automatically become an individual of all the superclasses of the class too.

The relationship between individuals can be expressed by simply using the property as the predicate in a statement, for example, using the `basedOn` property to describe the relationship between a depiction and the novel on which it is based (see Listing 2.23).

**Listing 2.23** Connecting individuals

```
@prefix ex: <http://example.com/> .
@prefix movie: <http://example.com/movie/> .
@prefix novel: <http://example.com/novel/> .

movie:DisneysZorro ex:basedOn novel:TheCurseOfCapistrano .
```

The *equality* and *inequality* of individuals can be expressed using `owl:sameAs` and `owl:differentFrom` (see Listing 2.24).

**Listing 2.24** Declaring identical and different individuals

```
@prefix movie: <http://example.com/movie/> .
@prefix owl: <http://www.w3.org/2002/07/owl#> .

movie:DonDiegoDeLaVega owl:sameAs movie:Zorro .
movie:CarlosMartinez owl:differentFrom movie:Zorro .
```

The individuals that belong to the same class can be enumerated using `owl:oneOf` (see Listing 2.25).

**Listing 2.25** Enumerating instances of a class

```
@prefix movie: <http://example.com/movie/> .
@prefix rdf: <http://www.w3.org/1999/02/22-rdf-syntax-ns#> .
@prefix owl: <http://www.w3.org/2002/07/owl#> .

movie:Villain a owl:Class ;
owl:equivalentClass [ a owl:Class ;
owl:oneOf ( movie:Monastario movie:Magistrado movie:Eagle ) ] .
```

## 2.5.3 Serialization

While all the previous examples were written in Turtle, OWL supports other serializations as well. OWL ontologies correspond to RDF graphs, i.e., sets of RDF triples. Similar to RDF graphs, OWL ontology graphs can be expressed in various syntactic notations. For the first version of OWL two special syntaxes were developed: the *XML presentation syntax*[17] and the *abstract syntax*.[18] Beyond the

---

[17]https://www.w3.org/TR/owl-xmlsyntax/

[18]https://www.w3.org/TR/2003/WD-owl-semantics-20030331/syntax.html

*RDF/XML syntax,*[19] the normative exchange syntax of both OWL and OWL 2, and
the *Turtle syntax* used in the previous examples, OWL 2 also supports *OWL/XML,*[20]
the *functional syntax,*[21] and the *Manchester syntax.*[22] To compare these syntaxes, a
code snippet of an ontology is serialized in each syntax, defining an ontology file,
prefixes for the namespaces, and the intersection of two classes. Listing 2.26 shows
the RDF/XML serialization.

**Listing 2.26** RDF/XML syntax example

```
<?xml version="1.0" encoding="UTF-8"?>

<rdf:RDF xml:base="http://vidont.org/"
 xmlns:owl="http://www.w3.org/2002/07/owl#"
 xmlns:rdf="http://www.w3.org/1999/02/22-rdf-syntax-ns#">

  <owl:Ontology rdf:about="vidont.ttl" />

  <owl:Class rdf:about="CartoonWithMotionPicture">
    <owl:equivalentClass>
      <owl:Class>
        <owl:intersectionOf rdf:parseType="Collection">
          <owl:Class rdf:about="Cartoon"/>
          <owl:Class rdf:about="LiveAction"/>
        </owl:intersectionOf>
      </owl:Class>
    </owl:equivalentClass>
  </owl:Class>

</rdf:RDF>
```

RDF/XML is the only syntax that all OWL 2-compliant tools must support;
however, it is rather verbose for big ontologies and somewhat difficult to read for
humans. This is why the more compact and easier-to-read RDF syntax, Turtle,
is commonly used for representing RDF triples for OWL 2 ontologies, as shown
in Listing 2.27.

**Listing 2.27** OWL/Turtle example

```
@prefix : <http://vidont.org/> .
@prefix owl: <http://www.w3.org/2002/07/owl#> .
```

---

[19]https://www.w3.org/TR/owl-mapping-to-rdf/

[20]https://www.w3.org/TR/owl-xml-serialization/

[21]https://www.w3.org/TR/owl-syntax/

[22]https://www.w3.org/TR/owl2-manchester-syntax/

```
@prefix rdf: <http://www.w3.org/1999/02/22-rdf-syntax-ns#> .

<http://vidont.org/vidont.ttl> a owl:Ontology .

:CartoonWithMotionPicture owl:equivalentClass [ a owl:Class ;
owl:intersectionOf ( :Cartoon :LiveAction ) ] .
```

OWL/XML is an alternate exchange syntax for OWL 2 (see Listing 2.28).

**Listing 2.28** OWL/XML example

```
<?xml version="1.0" encoding="UTF-8"?>
<Ontology
  xml:base="http://vidont.org/"
  ontologyIRI="http://vidont.org/vidont.ttl"
  xmlns="http://www.w3.org/2002/07/owl#">
<Prefix name="owl" IRI="http://www.w3.org/2002/07/owl#" />

  <EquivalentClasses>
    <Class IRI="CartoonWithMotionPicture" />
    <ObjectIntersectionOf>
      <Class IRI="Cartoon" />
      <Class IRI="LiveAction" />
    </ObjectIntersectionOf>
  </EquivalentClasses>

</Ontology>
```

OWL/XML was designed for XML tools that use XQuery, XPath, and XSLT, for which RDF/XML would be difficult to process (beyond parsing and rendering).

The OWL 2 functional syntax is clean, adjustable, modifiable, and easy to parse (see Listing 2.29).

**Listing 2.29** OWL 2 functional syntax example

```
Prefix(:=<http://vidont.org/>)
Prefix(owl:=<http://www.w3.org/2002/07/owl#>)

Ontology(<http://vidont.org/vidont.ttl>

  EquivalentClasses(
    :CartoonWithMotionPicture
    ObjectIntersectionOf( :Cartoon :LiveAction )
  )
)
```

The functional syntax makes the formal structure of ontologies clear, and it is almost exclusively used for defining the formal OWL 2 grammar in the high-level structural specifications of the W3C. The OWL 2 functional syntax is compatible with UML, one of the most widely deployed general-purpose standardized modeling languages.

The less frequently used Manchester syntax is a compact, user-friendly syntax for OWL 2 DL that collects information about a particular class, property, or individual into a single construct called a frame. The Manchester syntax is easy to read and write, especially for those who are not experts in mathematical logic. Complex descriptions consist of short, meaningful English words, while eliminating the logical symbols and precedence rules represented in other syntaxes, as shown in Listing 2.30.

**Listing 2.30**  Manchester syntax example

```
Prefix: : <http://vidont.org/>
Prefix: owl: <http://www.w3.org/2002/07/owl#>

Ontology: <http://vidont.org/vidont.ttl>

Class: CartoonWithMotionPicture
EquivalentTo: Cartoon and LiveAction
```

## 2.6  Simple Knowledge Organization System

A de facto standard for the formal representation of knowledge organization systems is called *Simple Knowledge Organization System (SKOS)*, an OWL ontology specially designed for creating controlled vocabularies, classification schemes, taxonomies, subject headings, and thesauri [46]. The vocabulary of SKOS provides terms for basic structure (skos:Concept), lexical labeling (skos:prefLabel, skos:altLabel), documentation (skos:note, skos:definition, skos:example, skos:scopeNote, skos:historyNote), semantic relations (skos:broader, skos:narrower, skos:related), concept schemas (skos:ConceptScheme, skos:hasTopConcept), subject indexing (skos:subject), and node labels (skos:Collection, skos:member). A short demonstration of how the SKOS terms are used to define concepts, their preferred and alternate labels, along with broader, narrower, similar, or related concepts is shown in Listing 2.31.

**Listing 2.31**  Demonstration of core SKOS terms

```
@prefix ex: <http://example.com/> .
@prefix rdf: <http://www.w3.org/1999/02/22-rdf-syntax-ns#> .
```

```
@prefix skos: <http://www.w3.org/2004/02/skos/core#> .

ex:RedKangaroo a skos:Concept ; skos:prefLabel "Red Kangaroo"
; skos:altLabel "Osphranter rufus" ;
skos:broader ex:AustralianMarsupial ; skos:related ex:Wallaby .
```

Note that while it seems logical to have a broader term in the subject than in the object when using skos:broader, the SKOS primer specification uses skos:broader the other way around: the broader term follows the narrower one in the RDF triple, and hence the predicate should read as "has broader concept" rather than "broader than" [47]. Similarly, the subject should be broader than the object in RDF triples using the skos:narrower predicate.

Knowledge organization systems utilizing SKOS can combine SKOS terms with properties and relations from other vocabularies and ontologies, for example, from rdfsV (e.g., rdfs:partOf), OWL (e.g., owl:subClassOf, owl:instanceOf), and Dublin Core (e.g., dc:title, dc:description, dc:rights, dc:creator). While some argue, or are confused about, the difference between certain SKOS and OWL relations (e.g., skos:Concept and owl:Class, or skos:exactMatch and owl:sameAs), and might convert SKOS-based taxonomies and thesauri to OWL [48], there are metamodeling differences between SKOS and OWL. SKOS concepts correspond to OWL Full individuals, but support class-level characteristics as well [49].[23] While owl:sameAs links an individual to an individual, indicating that two IRI references actually refer to the same resource, the skos:exactMatch property links two concepts that are sufficiently similar to be used interchangeably in a particular application, but RDF triples containing the first resource do not necessarily constitute a true statement when the second resource is substituted in.

SKOS is widely used in the IT industry, and implementation examples include not only large-scale subject headings and thesauri, but also semantic Big Data (Big Semantic Data) applications powered by knowledge formalisms. Since 2009, the New York Times subject headings are available as SKOS.[24] Since 2012, the ACM Computing Classification System, the de facto standard for classifying computer science literature, is available in SKOS [50]. The largest library in the world, the Library of Congress, utilizes SKOS in its subject headings and thesauri.[25] The Food and Agriculture Organization of the United Nations (FAO) provides a SKOS-aligned version of their thesaurus, AGROVOC, which contains more than 40,000 concepts.[26]

---

[23]In OWL (2) DL, the sets of classes and individuals are disjoint, therefore SKOS concepts cannot be treated as OWL (2) DL classes.

[24]https://datahub.io/dataset/nytimes-linked-open-data

[25]http://id.loc.gov/download/

[26]http://aims.fao.org/standards/agrovoc/linked-open-data

Some specific software tools that support SKOS editing are the SKOSEd Protégé plugin,[27] Enterprise Vocabulary Net,[28] and PoolParty.[29]

## 2.7   Rule Languages

The information that enables software agents to make new discoveries is based on RDF statements and *rulesets* (collections of *rules*, i.e., IF-THEN constructs). While ontologies focus on the classification methods by defining classes, sub-classes, and relations, rulesets focus on general mechanisms for discovering and generating new relations based on existing relations. If the condition in the IF part of the code holds, the conclusion of the THEN part of the code is processed. Rules are relatively easy to implement, and can express existential quantification, disjunction, logical conjunction, negation, functions, nonmonotonicity, and other features.

The two most popular rule languages on the Semantic Web are the *Rule Markup Language (RuleML)*, an XML approach to represent both forward (bottom-up) and backward (top-down) rules,[30] and the *Semantic Web Rule Language (SWRL)*, which was introduced as an extension to OWL, and combines OWL and RuleML [51].

### 2.7.1   Semantic Web Rule Language

The conceptualization of the knowledge domain related to an image or a video clip may be covered by one or more multimedia ontologies, which can be used to formally represent depicted concepts, their properties, and the relationships between them. However, even OWL 2 DL is not expressive enough to describe all kinds of multimedia contents, such as complex video events. There are decidable rule-based formalisms, such as function-free *Horn rules*, which are not restricted in this regard.

While some OWL 2 axioms, such as class inclusion and property inclusion, correspond to rules, and other classes can be decomposed to rules, and property chain axioms provide rule-like axioms, there are rule formalisms that cannot be expressed in OWL 2. For example, a rule head with two variables cannot be represented as a subclass axiom, or a rule body that contains a class expression cannot be described by a subproperty axiom. To add the additional expressivity of rules to OWL 2 DL, ontologies can be extended with SWRL rules, which are

---

[27]https://code.google.com/archive/p/skoseditor/

[28]http://www.topquadrant.com/products/topbraid-enterprise-vocabulary-net/

[29]https://www.poolparty.biz/skos-and-skos-xl/

[30]http://www.ruleml.org

expressed in terms of OWL classes, properties, and individuals. SWRL rules are datalog-style rules with unary predicates for describing classes and data types, binary predicates for properties, and some special built-in n-ary predicates (see Definition 2.16). SWRL rules contain an antecedent (*body*) and a consequent (*head*), both of which can be an atom or a positive (i.e., unnegated) conjunction of atoms (see Definition 2.15).[31]

**Definition 2.15 (Atom)** An *atom* is an expression of the form $P(\text{arg}_1, \text{arg}_2, \ldots)$, where $P$ is a predicate symbol (classes, properties, or individuals) and $\text{arg}_1, \text{arg}_2, \ldots$ are the arguments of the expression (individuals, data values, or variables).

**Definition 2.16 (SWRL Rule)** A *rule* R is given as $B_1 \wedge \ldots \wedge B_m \rightarrow H_1 \wedge \ldots \wedge H_n$ $(n \geqslant 0, m \geqslant 0)$, where $B_1, \ldots, B_m, H_1, \ldots H_n$ are atoms, $B_1 \wedge \ldots \wedge B_m$ is called the body (premise or antecedent), and $H_1 \wedge \ldots \wedge H_n$ is the head (conclusion or consequent).

The following SWRL atom types exist:

- *Class atom*: an OWL named class or class expression and a single argument representing an OWL individual, e.g., `Movie(?m)`
- *Individual property atom*: an OWL object property and two arguments representing OWL individuals, e.g., `hasEpisode(?x, ?y)`
- *Data range-restricted property atom*: an OWL data property and two arguments (OWL individual, data value), e.g., `hasDuration(?x, ?duration)`
- *Atom of different or same individuals* (`differentFrom`, `sameAs`): an atom to determine whether two individuals refer to the same individual or are different, e.g., `sameAs(?x, ?y)`
- *Built-in atom*: a predicate that takes one or more arguments and evaluates to true if the arguments satisfy the predicate, e.g., `swrlb:greaterThan(?age, 17)`

  - SWRL built-ins for comparisons: `equal`, `notEqual`, `lessThan`, `lessThanOrEqual`, `greaterThan`, `greaterThanOrEqual`
  - SWRL built-ins for Math: `add`, `subtract`, `multiply`, `divide`, `integerDivide`, `mod`, `pow`, `unaryPlus`, `unaryMinus`, `abs`, `ceiling`, `floor`, `round`, `roundHalfToEven`, `sin`, `cos`, `tan`
  - SWRL built-in for Boolean values: `booleanNot`
  - SWRL built-ins for strings: `contains`, `containsIgnoreCase`, `endsWith`, `lowerCase`, `matches`, `normalizeSpace`, `replace`, `startsWith`, `stringConcat`, `stringEqualIgnoreCase`, `stringLength`, `substring`, `substringAfter`, `substringBefore`, `tokenize`, `translate`, `uppercase`
  - SWRL built-ins for date, time, and duration: `addDayTimeDurationToDate`, `addDayTimeDurationToDateTime`, `addDayTimeDurationToTime`, `addDayTimeDurations`,

---

[31]Since SWRL is monotonic, it does not support negated atoms, retraction, and counting.

addYearMonthDurationToDate,
addYearMonthDurationToDateTime, addYearMonthDurations,
date, dateTime, dayTimeDuration, divideDayTimeDuration,
divideYearMonthDurations, multiplyDayTimeDurations,
multiplyYearMonthDuration,
subtractDateTimesYieldingDayTimeDuration,
subtractDateTimesYieldingYearMonthDuration,
subtractDates, subtractDayTimeDurationFromDate,
subtractDayTimeDurationFromDateTime,
subtractDayTimeDurationFromTime,
subtractDayTimeDurations, subtractTimes,
subtractYearMonthDurationFromDate,
subtractYearMonthDurationFromDateTime,
subtractYearMonthDurations, time, yearMonthDuration

- Temporal SWRL built-ins: add, after, before, contains,
  duration, durationEqualTo, durationGreaterThan,
  durationLessThan, during, equals, finishedBy, finishes,
  meets, metBy, notAfter, notBefore, notContains,
  notDurationEqualTo, notDurationGreaterThan,
  notDurationLessThan, notDuring, notEquals, notFinishedBy,
  notFinishes, notMeets, notMetBy, notOverlappedBy,
  notOverlaps, notStartedBy, notStarts, now, overlappedBy,
  overlaps, startedBy, starts
- SWRL built-ins for URIs: anyURI, resolveURI
- SWRL built-ins for lists: empty, first, length, listConcat,
  listIntersection, listSubtraction, member, rest, sublist
- Custom SWRL built-ins

The typical prefix for SWRL is swrl: and the namespace URI is http://
www.w3.org/2003/11/swrl#. The prefix for SWRL built-ins is swrlb:
and the namespace URI is http://www.w3.org/2003/11/swrlb#. SWRL
rule editors often omit the namespace declarations.

Note that SWRL rules can be written either as implication or reverse implication. The direction of the arrow in a SWRL rule indicates when the rule is fired. The
left-pointing arrow ($\rightarrow$) appears in *forward rules* that fire on rule addition and new
statement addition. The right-pointing arrow ($\leftarrow$) is used in *backward rules* that fire
during query expansion. For example, the SWRL rule for describing that the brother
of someone's parent is his uncle can be written as shown in Listing 2.32.

**Listing 2.32** The "has uncle" relationship in SWRL

```
hasParent(?x, ?y) ^ hasBrother(?y, ?z)  →  hasUncle(?x, ?z)
```

The first version of OWL was not expressive enough to describe such relationships, which was addressed in OWL 2 with the introduction of chained properties
(see later in Chap. 4).

To express that John's female sibling is his sister, one can use the named
individual "John" in the SWRL rule (see Listing 2.33).

**Listing 2.33** The "has sister" relationship for the named individual "John"

```
Person(John) ^ hasSibling(John, ?s) ^ Woman(?s) →
hasSister(John, ?s)
```

The SWRL rule in Listing 2.34 determines whether a person is a child or an adult
based on his age using the lessThan SWRL built-in.

**Listing 2.34** A SWRL rule using the lessThan comparison built-in

```
Person(?p) ^ hasAge(?p,?age) ^ swrlb:lessThan(?age,18) →
Child(?p)
```

SWRL rules can also be used for mathematical calculations on numerical data,
such as to determine the price of a product in different currencies based on the
current exchange rate (see Listing 2.35).

**Listing 2.35** A SWRL rule with built-in argument binding

```
Product(?p) ^ hasPriceInAUD(?p, ?aud) ^ swrlb:multiply(?usd,
?aud, ?usdaudexchangerate) -> hasPriceInUSD(?p, ?usd)
```

String built-ins are useful for string manipulation tasks, such as concatenation,
determining string length, checking whether a string matches a regular expression,
etc. Listing 2.36 employs the startsWith SWRL built-in to verify whether the
phone number of an enterprise is an international phone number.

**Listing 2.36** A SWRL rule using the startsWith string built-in

```
Enterprise(?e) ^ hasPhoneNumber(?p, ?phone) ^
swrlb:startsWith(?phone, "+") →
hasInternationalPhoneNumber(?e, true)
```

One of the main benefits of SWRL is that it is a declarative rule language
independent from the execution algorithm. In contrast, all other rule languages,
such as Jess, RIF-PRD (Rule Interchange Format Production Rule Dialect), or the
ILOG JRules used by IBM WebSphere applications, are bound to a specific (usually
Rete-based) execution algorithm. Consequently, in SWRL the most efficient evaluation algorithm can be implemented for any given domain, and this algorithm can
be changed any time without affecting the ruleset. Also, SWRL rules written for a
particular application can often be reused in different scenarios. Based on model-

theoretic semantics that can be seamlessly integrated with OWL, SWRL rules do not require an intermediate language the rule engine can interpret (as opposed to Jess, JRules, or RIF-PRD, for example). Being similar to the semantics of PRO-LOG, SWRL semantics allow the use of well-established declarative debugging techniques.

### 2.7.2 Rule Interchange Format

Due to the different paradigms, semantics, features, syntaxes, and target audience of rule languages, there is a need for rule exchange. The *Rule Interchange Format (RIF)* was designed for rule sharing and exchange between existing rule systems by allowing rules written for one application to be shared and reused in other applications and rule engines while preserving semantics. RIF is a collection of rigorously defined rule languages called *dialects*. The *Core Dialect* of RIF is a common subset of most rule engines. The *Basic Logic Dialect (BLD)* adds logic functions, equality and named arguments, and supports Horn logic. The *Production Rules Dialect (PRD)* is an extension of the Core Dialect, and defines a language for production rules, written in the form of condition-action rules, which are used to specify behaviors and separate business logic from business objects.

## 2.8 Structured Data Deployment

RDF triples, which can use arbitrary terms from RDFS vocabularies and OWL ontologies, are typically serialized in structured datasets and purpose-built graph databases in a mainstream RDF serialization format, such as RDF/XML or Turtle. RDF triples can also be deployed as lightweight annotations embedded in the website markup in RDFa, HTML5 Microdata, and JSON-LD.

### 2.8.1 Linked Open Data Datasets

In contrast to the isolated data silos of the conventional World Wide Web, the Semantic Web interconnects open data, so that all datasets contribute to a global data integration, connecting data from diverse domains, such as people, companies, books, journal papers, scientific data, medicine, statistics, films, music, reviews, television and radio programs, and online communities. The union of the structured datasets forms the *Linked Open Data Cloud (LOD Cloud)*,[32] the decentralized core of the Semantic Web, where software agents can automatically find relationships

---

[32]http://lod-cloud.net

between entities and make new discoveries. Linked Data browsers allow users to browse a data source, and by using special (typed) links, navigate along links into other related data sources. Linked Data search engines crawl the *Web of Data* by following links between data sources to provide expressive query capabilities over aggregated data. The intelligent automation enabled by *Linked Open Data (LOD)* is universally exploited by search engines, governments, social media, publishing agencies, media portals, researchers, and individuals.

### 2.8.1.1  Linked Data Principles

Conventional web pages are hypertext documents connected with hyperlinks (or simply links). These hyperlinks point to other documents or a part of another document; however, they do not hold information about the type of relationship between the source and the destination resources. While the link relation can be annotated using the `rel` attribute on the `link`, `a`, and `area` markup elements, the `rel` attribute is suitable for annotating external CSS files, script files, or the favicon only. On the Semantic Web, links can be typed by using the `rdf:type` predicate and providing a machine-interpretable definition for the relationship between the source and destination resources. *Linked Data* refers to the best practices for publishing structured data to be interlinked with typed links, thereby enabling semantic querying across diverse resources [52]. The term was coined by Tim Berners-Lee, who defined the four principles of Linked Data, which can be summarized as follows [53]:

1. Use URIs as names for the representation of real-world concepts and entities, i.e., assign a dereferenceable URI to each resource rather than an application-specific identifier, such as a database key or incremental numbers, making every data entity individually identifiable.
2. Use HTTP URIs, so that resource names can be looked up on the Internet, i.e., provide URIs of RDF resources over the HTTP protocol for dereferencing.
3. The resource URIs should provide useful information using Semantic Web standards, such as RDF. For example, provenance data and licensing metadata about the published data can help clients assess data quality and choose between different means of access.
4. Include links to other URIs, so that users and software agents can discover related information. When RDF links are set to other data resources, users can navigate the Web of Data as a whole by following RDF links.

More and more organizations, businesses, and individuals are recognizing the benefits of Linked Data. Several industry giants that are already utilizing LOD include Amazon.com, the BBC, Facebook, Flickr, Google, Yahoo!, Thomson Reuters, and The New York Times.

### 2.8.1.2   The Five-Star Deployment Scheme for Linked Data

Publishing Linked Data in accordance with the four Linked Data principles does not guarantee data quality. For example, the documents with which the URIs of LOD datasets are interlinked might be difficult to reuse, such as a PDF file containing a table as a scanned image. A five-star rating system[33] is used to express the quality of Linked Data. The five-star rating system is cumulative, so that data have to meet all lower-level criteria before attaining a higher-level rating:

| | |
|---|---|
| ★ | Data available on the Web in any format, which is human-readable but not machine-interpretable, due to a vendor-specific file format or lack of structure. All the following stars are intended to make the data easier to discover, use, and understand. For example, a scanned image of tabular data in a PDF file is one-star data, because data reusability is limited. |
| ★★ | Data available as machine-readable structured data. For example, tabular data saved in an Excel file is two-star data. |
| ★★★ | Data available in a nonproprietary (vendor-independent) format. For example, tabular data saved as a CSV file is three-star data. |
| ★★★★ | Data published using open standards, such as RDF and SPARQL. For example, tabular data in HTML with RDFa annotation using URIs is four-star data. |
| ★★★★★ | Data that meets all of the above criteria and complemented by links to other, related data to provide context. For example, tabular data in the website markup semantically enriched with HTML5 Microdata is five-star data, because reusability and machine-interpretability are maximized. |

Open licensing enables free data reuse by granting permission to access, reuse, and redistribute a creative work, such as text, image, audio, or video, with few or no restrictions. Linked Data without an explicit open license (e.g., public domain license) cannot be reused freely, but the quality of Linked Data is independent from licensing. Since the five-star deployment scheme applies to both Linked Data (Linked Data without an open license)[34] and *Linked Open Data* (*LOD*, Linked Data with an open license), the five-star rating system can be depicted in a way that the criteria can be read with or without the open license.

Because converting a CSV file to a set of RDF triples and interlinking them with other sets of triples does not necessarily make the data more (re)usable to humans or software agents, even four-star and five-star Linked Open Data have many challenges. One of the challenges is the lack of provenance information, which can now be provided about Linked (Open) Data using the de facto standard *PROV-O* ontology.[35] Another challenge is querying Linked Data that do not use machine-readable definitions from a vocabulary or ontology, which is difficult, and sometimes infeasible, to interpret for software agents. Furthermore, the quality of the

---

[33]http://5stardata.info

[34]This licensing concept is used on the conventional Web too, where the term "Open Data" refers to the free license.

[35]https://www.w3.org/TR/prov-o/

definitions retrieved from vocabularies and ontologies varies greatly, and implementations do not always utilize the best possible classes and properties.

### 2.8.1.3  LOD Datasets

A meaningful collection of RDF triples covering a field of interest according to the Linked Open Data principles is called an *LOD dataset*. LOD datasets collect descriptions of entities within a field of interest, such as music (e.g., DBTune,[36] LinkedBrainz),[37] movies (e.g., LinkedMDB),[38] and encyclopedic content (e.g., DBpedia).[39] The authors of well-established datasets provide advanced features that enable easy access to their structured data, such as downloadable compressed files of the datasets or an infrastructure for efficient querying.

Similar to the web crawlers that systematically browse conventional websites for indexing, Semantic Web crawlers browse semantic contents to extract structured data and automatically find relationships between related entities (regardless that the relation is evident or not). All properly published LOD datasets are available through *RDF crawling*.

The RDF triples of the most popular LOD datasets are regularly published as a downloadable compressed file (usually gzip or bzip2), called an *RDF dump*. The reason behind compressing RDF dump files is the good compressibility of text and the relatively large file size (in the GB range). The size of gzip-compressed RDF dumps is typically around 100MB per every 10 million triples, but it also depends on data redundancy and the RDF serialization of the dataset.

Similar to relational database queries in SQL, the data of semantic datasets can be retrieved through powerful queries. The query language designed specially for RDF datasets is called *SPARQL* (pronounced "sparkle," it stands for *SPARQL Protocol and RDF Query Language*). Some datasets provide a *SPARQL endpoint*, which is an address from which SPARQL queries can be directly executed.

### 2.8.1.4  Licensing

Linked Open Data without an explicit license is just Linked Data. To make LOD datasets truly "open," the open license has to be declared explicitly, which prevents potential legal liability issues and makes it clear to users what usage conditions apply. The licensing information of a dataset can be provided in the dataset file or in an external metadata file, such as a VoID (Vocabulary of Interlinked Datasets) file. The license under which the dataset has been published can be declared using the

---

[36]http://dbtune.org

[37]http://linkedbrainz.c4dmpresents.org

[38]http://linkedmdb.org

[39]http://dbpedia.org

`dcterms:license` property. The most frequently used licenses for Linked Open Data are the following:

- Public Domain Dedication and License (PDDL)—"Public Domain for Data/ Databases"[40]
- Open Data Commons Attribution (ODC-By)—"Attribution for Data/Data- bases"[41]
- Open Database License (ODC-ODbL)—"Attribution Share-Alike for Data/ Databases"[42]
- CC0 1.0 Universal (CC0 1.0)—"Public Domain Dedication"[43]
- Creative Commons Attribution-ShareAlike (CC-BY-SA)[44]
- GNU Free Documentation License (GFDL)[45]

The first four licenses are specially designed for data, so their use for LOD dataset licensing is highly recommended. Licensing of datasets is a complex issue, because datasets are collections of facts rather than creative works, so different laws apply. Creative Commons and GPL are quite common on the Web; however, they are based on copyright and are designed for creative works, not datasets, so they might not have the desired legal result when applied to datasets.

Community norms (nonbinding conditions of use) can be expressed using the `waiver:norms` property (`http://vocab.org/waiver/terms/ norms`). A common community norm is ODC Attribution ShareAlike (`http://www.opendatacommons.org/norms/odc-by-sa/`), which permits data use from the dataset, but the changes and updates should be public too.

### 2.8.1.5   Interlinking

Government agencies, large enterprises, media institutes, social media portals, and researchers work with large amounts of data that can be represented as structured data and published as Linked Data. However, the structured representation of government data, university departments, and research groups is often produced as an isolated dataset file, which is not part of the Semantic Web until it is linked to LOD datasets.

Creating links between the structured datasets of the Semantic Web is called *interlinking*, which makes isolated datasets part of the LOD Cloud, in which all resources are linked to one another. These links enable semantic agents to navigate between data sources and discover additional resources. The most common

---

[40]https://www.opendatacommons.org/licenses/pddl/

[41]https://www.opendatacommons.org/licenses/by/

[42]https://www.opendatacommons.org/licenses/odbl/

[43]https://creativecommons.org/publicdomain/zero/1.0/

[44]https://creativecommons.org/licenses/by-sa/4.0/

[45]http://www.gnu.org/copyleft/fdl.html

predicates for interlinking are `owl:sameAs` and `rdfs:seeAlso`, but any predicate can be used. In contrast to hyperlinks between web pages, LOD links utilize typed RDF links between resources. The URIs of the subject and the object of the link identify the interlinked resources. The URI of the predicate defines the type of the link. For example, an RDF link can state that a person is employed by a company, while another RDF link can state that the person knows other people. Dereferencing the URI of the link destination yields a description of the linked resource, usually containing additional RDF links that point to other, related URIs, which, in turn, can also be dereferenced, and so on.

## 2.8.2   Graph Databases: Triplestores and Quadstores

To leverage the power of RDF, data on the Semantic Web can be stored in *graph databases* rather than relational databases. Graph databases store RDF triples and exploit graph structures to perform semantic queries using nodes, edges, and properties that represent structured data. Graph databases implement Create, Read, Update, and Delete (CRUD) methods while exposing a graph data model. These include property graphs that contain nodes and relationships, hypergraphs in which a relationship can connect an arbitrary number of nodes, RDF triples, or RDF quads. A *quad* is a subject-predicate-object triple coupled with a graph identifier (see Definition 2.17). The identified graph is called a *named graph*, which is a pragmatic alternative to RDF reification.[46]

**Definition 2.17 (RDF Quadruple)** Let $S$ be a set of data sources, which is a subset of the set of IRIs, i.e., $S \subset \mathbb{I}$. An ordered pair $(t, g)$ of a triple $t := (s, p, o)$ and a graph $g \in S$ and $t \in \text{get}(c)$ is a *triple in graph g*, where get is a HTTP `get` request, and the result of the request cannot be an empty set, i.e., $\text{get}(s) \neq \emptyset$. The quadruple $(s, p, o, g)$ is called an *RDF quadruple (RDF quad)*.

A graph database that stores RDF triples is called a *triplestore* (subject-predicate-object database). A graph database that stores the graph name (representing the graph context or provenance information) for each triple is called a *quadstore*. Quadstores store the graph URI along with all the triples, so that all RDF statements become globally unique and dereferenceable. This can be used, for example, to make statements about entire RDF graphs or provide provenance data for RDF triples.

Most graph databases can be used with online transactional processing (OLTP) systems and are optimized for high performance and integrity. In contrast to relational and NoSQL databases, purpose-build triplestores and quadstores do not

---

[46]Compared to reification, named graphs avoid triple bloat and eliminate the problems associated with reified triple querying and metadata redundancy.

depend on indices, because graphs provide an adjacency index by default, and relationships assigned to nodes provide a direct connection to other, related nodes. This enables graph queries to traverse through the graph, which is usually far more efficient than querying relational databases that join data using a global index. In fact, many graph databases are so powerful that they are utilized in Big Data applications.

Many benefits of graph databases are derived from the Semantic Web standards they utilize. The globally unique and interlinked RDF subjects exploit the advantages of the graph structure. Adding a new schema element is as simple as inserting a triple with a new predicate. Graph databases also support ad hoc SPARQL queries. In contrast to the column headers, foreign keys, or constraints typical to relational databases, graph database entities are categorized using classes and predicates that correspond to properties or relationships, which are all parts of the data. Graph databases support not only manual and programmatic querying, but also automated reasoning for knowledge discovery.

## 2.8.3  Lightweight Annotations

The three RDF serializations that can directly be embedded into the website markup are the following:[47]

- *RDFa*, which expresses RDF in markup attributes that are not part of the core (X)HTML vocabularies
- *HTML5 Microdata*, which extends the HTML5 markup with structured metadata (an HTML5 API)
- *JSON-LD*, which adds structured data to the markup as JavaScript code

RDFa ("RDF in attributes") is a lightweight annotation for writing RDF triples directly in host languages, such as (X)HTML, XML, and SVG, as attribute values. The full RDFa syntax (*RDFa Core*) [54] provides advanced features to express arbitrary RDF statements as structured data in the markup. A minimal subset of RDFa is the less expressive *RDFa Lite* [55], which is easier to learn and suitable for most general scenarios. The attributes supported by RDFa Lite include `vocab`, `typeof`, `property`, `resource`, and `prefix`. The `vocab` attribute provides the option to declare a single vocabulary for part of the markup code. The `typeof` attribute is a whitespace-separated list of IRIs that indicates the RDF type(s) to associate with a subject. The `property` attribute is used for expressing predicates, such as to describe the relationship between a subject and a resource object or string literal. The `resource` attribute is used to set the object of statements. The `prefix`

---

[47]The first set of lightweight structured annotation formats, microformats, was used as core XHTML attribute values (rather than defining an own syntax), but was limited to basic concepts, such as people, places, events, recipes, and audio, hence microformats are omitted here.

attribute assigns prefixes to represent URLs and, using those prefixes, the vocabulary elements can be abbreviated as usual. RDFa Lite annotations can also implement the `href` attributes, which express the resource object with a navigable IRI,
and `src` attributes, which express the resource object for embedded resources, as
long as the host language supports them.

More sophisticated annotations require additional attributes that are supported
by RDFa Core only. Beyond the RDFa Lite attributes, RDFa Core supports the
`about`, `content`, `datatype`, `inlist`, `rel`, and `rev` attributes. Without
declaring the subject of RDF triples, the subject of all the RDF statements in a
document will be the IRI of the document being parsed. In other words, by default
all RDF triples will make statements about the document itself. In markup languages, this can be overridden by using the `base` element in order to make
statements about a particular resource, or using the `about` attribute to change
the subject of a single statement. Objects that are plain literals can be declared using
the `content` attribute. Typed literals can be given by declaring a datatype using
the `datatype` attribute.

In RDFa, the relationship between two resources (i.e., the predicates) can
also be expressed using the `rel` attribute. When a predicate is expressed using
`rel`, the `href` or `src` attribute applied on the element of the RDFa statement
identifies the object. Reverse relationships between two resources can be expressed
with the `rev` attribute. The `rel` and `rev` attributes can be used on any element
individually or together. Combining `rel` and `rev` is particularly useful when
there are two different relationships to express, such as when a photo is taken
by the person it depicts. Note that annotating a predicate with `rel` or `rev`
only (without `href`, `src`, or `resource`) will result in an RDF triple with a
blank node (incomplete triple) [56]. The `inlist` attribute indicates that the object
generated on the element is part of a list sharing the same predicate and subject.
Only the presence of the `inlist` attribute is relevant; its attribute value is always
ignored.

HTML5 Microdata is an HTML5 module defined in a separate specification,
extending the HTML5 core vocabulary with attributes for representing structured
data [57]. HTML5 Microdata represents structured data as a group of name-value
pairs. The groups are called items, and each name-value pair is a property. Items
and properties are represented by regular (X)HTML5 elements. To create an item,
the `itemscope` attribute is used.[48] To add a property to an item, the `itemprop`
attribute is used on a descendant of the item (a child element of the container

---

[48]In HTML5, it is a common practice to use attribute minimization and omit the attribute value,
which is not allowed in XHTML5. For example, in HTML5, `itemscope` can be written on the
container element without a value, while XHTML5 only allows `itemscope="itemscope"`,
which is more verbose but also more precise, and validates both as HTML5 and XHTML5
(polyglot markup).

element). Groups of name-value pairs can be nested in a Microdata property by declaring the itemscope attribute on the element that declared the property. Property values are usually string literals, but can also be IRIs as the value of the href attribute on the a element, the value of the src attribute on the img element, or other elements that link to or embed external resources. The type of the items and item properties are expressed using the itemtype attribute, by declaring the URL of the external vocabulary that defines the corresponding items and properties. If an item has a global identifier (such as the barcode number of a Blu-Ray release), it can be annotated using the itemid attribute. Elements with an itemscope attribute may have an optional itemref attribute, which gives a list of additional elements to crawl to find the name-value pairs of the item. In other words, properties that are not descendants of the element with the itemscope attribute can be associated with the item using the itemref attribute, providing a list of element identifiers with additional properties elsewhere in the document. The itemref attribute is not part of the Microdata data model, and is purely a syntactic construct to annotate web page components for which creating a tree structure is not straightforward, as, for example, a table in which the columns represent items and the cells the properties.

In contrast to RDFa and HTML5 Microdata, *JavaScript Object Notation for Linked Data (JSON-LD)*, the other mainstream format to add structured data to website markup, is written in JavaScript code blocks rather than markup elements and attributes. Consequently, JSON-LD is separate from the (X)HTML code. One of the advantages of this lightweight Linked Data format is that it is easy for humans to read and write. JSON-LD transports Linked Data using the JavaScript Object Notation (JSON), an open standard format using human-readable text to transmit attribute-value pairs [58]. If the JSON-LD code is written in a separate file rather than the markup, the de facto file extension is .jsonld. The Internet media type of JSON-LD is application/ld+json and, if written in the markup, the JSON-LD code is delimited by curly brackets (braces) between the <script> and </script> tags.

RDFa and JSON-LD can be used in most markup language versions and variants, while HTML5 Microdata can only be used in (X)HTML5. All these annotation formats have their own syntax. For example, the vocab attribute of RDFa is equivalent to the itemtype attribute in Microdata and @context in JSON-LD, all of which declare the vocabulary (Table 2.1).

All the previous serializations are the machine-interpretable equivalents of the natural language sentence "ABC is broadcasting Australian Story at 8pm on June 13, 2016."

**Table 2.1**  Structured data in different formats

| Annotation format | Structured data |
|---|---|
| Markup with RDFa | ```<div vocab="http://schema.org"``` <br> ```typeof="BroadcastEvent">``` <br>     ```<span property="name">Australian Story -``` <br>     ```S2016E20 -20th Anniversary Special (Part``` <br>     ```2)</span>``` <br>     ```<span property="description">This two-part``` <br>     ```anniversary special looks back over Australian``` <br>     ```Story's key themes.</span>``` <br>     ```<span property="startDate" content="2016-``` <br>     ```06-13T20:00">starts 13/06/2016 8pm</span>``` <br>     ```<span property="endDate" content="2016-06-``` <br>     ```13T20:29">ends 13/06/2016 8:29pm</span>``` <br>   ```<div property="publishedOn"``` <br>   ```typeof="BroadcastService">``` <br>     ```<span property="name">ABC</span>``` <br>   ```</div>``` <br> ```</div>``` |
| Markup with HTML5 Microdata | ```<div itemscope``` <br> ```itemtype="http://schema.org/BroadcastEvent">``` <br>   ```<span itemprop="name">Australian Story -``` <br>   ```S2016E20 - 20th Anniversary Special (Part``` <br>   ```2)</span>``` <br>   ```<span itemprop="description">This two-part``` <br>   ```anniversary special looks back over Australian``` <br>   ```Story's key themes.</span>``` <br>   ```<span itemprop="startDate" content="2016-06-``` <br>   ```13T20:00">starts 13/06/2016 8pm</span>``` <br>   ```<span itemprop="endDate" content="2016-06-``` <br>   ```13T20:29">ends 13/06/2016 8:29pm</span>``` <br>   ```<div itemprop="publishedOn" itemscope``` <br>   ```itemtype="http://schema.org/BroadcastService">``` <br>     ```<span itemprop="name">ABC</span>``` <br>   ```</div>``` <br> ```</div>``` |

(continued)

**Table 2.1**   (continued)

| Annotation format | Structured data |
|---|---|
| Markup with JSON-LD | ```<script type="application/ld+json">``` <br> {... } |

```
<script type="application/ld+json">
  {
     "@context":"http://schema.org",
     "@type":"BroadcastEvent",
     "startDate":"2016-06-13T20:00",
     "endDate":"2016-06-13T20:29",
     "publishedOn": {
       "@type":"BroadcastService",
       "name: "ABC"
     },
     "name":"Australian Story - S2016E20 - 20th
     Anniversary Special (Part 2)",
     "description":"This two-part anniversary
     special looks back over Australian Story's key
     themes."
  }
</script>
```

## 2.9   Summary

This chapter has explained the fundamental concepts of knowledge representation and shown the difference between unstructured, semistructured, and structured data. It described core Semantic Web standards, such as RDF, the cornerstone of the Semantic Web; RDFS, the extension of the vocabulary of RDF for creating vocabularies, thesauri, and basic ontologies; OWL, the fully featured de facto standard ontology language; and SWRL, the Semantic Web Rule Language.

# Chapter 3
# The Semantic Gap

With the constantly increasing number of videos shared online, the automated processing of video files remains a challenge, mainly due to the huge gap between what computers can interpret and what humans understand, known as the *Semantic Gap*. Automatically extractable low-level features, such as dominant color or color distribution, are suitable for a limited range of practical applications only and are not connected directly to sophisticated human-interpretable, high-level descriptors, which can be added manually. Several attempts have been made in the last decade to bridge the semantic gap by mapping semistructured controlled vocabularies, in particular the vocabulary of MPEG-7, to RDFS and OWL. However, these mappings inherited issues from the original XML or XSD vocabularies, some of which can be addressed by combining upper and domain ontologies, rule formalisms, and information fusion.

## 3.1 Low-Level Descriptors

Low-level descriptors describe automatically extractable low-level image, audio, and video features, i.e., local and global characteristics of images and audio and video signals, such as intensity, frequency, distribution, pixel groups, and low-level feature aggregates, such as various histograms and moments based on low-level features. Such descriptors, many of which are defined in the MPEG-7 standard, are suitable for the numeric representation of audio waveforms as well as edges, interest points, regions of interest, ridges, and other visual features of images and videos. Well-established implementations of low-level feature extraction algorithms are available in the form of software libraries (e.g., OpenCV),[1] MATLAB

---

[1]http://opencv.org

© Springer International Publishing AG 2017
L.F. Sikos, *Description Logics in Multimedia Reasoning*,
DOI 10.1007/978-3-319-54066-5_3

functions (Computer Vision System Toolbox),[2] etc., and so software engineers can
implement low-level feature extraction in their programs without developing feature extractors from scratch.

The following sections provide a quick overview of the most common low-level
visual and auditory descriptors used for image, audio, and video annotation and
content-based retrieval.

### 3.1.1  Common Visual Descriptors

The most common perceptual categories of visual descriptors for still images are
color, texture, and shape. Image sequences add one more dimension of perceptual
saliency to these: motion.

- **Color descriptors** The *dominant color descriptor* specifies a set of dominant
  colors for an image (typically 4–6 colors), and considers the percentage of image
  pixels each color is used in, as well as the variance and spatial coherence of the
  colors. The *color structure descriptor* encodes local color structure by utilizing a
  structuring element, visiting all locations in an image, and summarizing the
  frequency of color occurrences in each structuring element. The *color layout
  descriptor* represents the spatial distribution of colors. The *scalable color
  descriptor* is a compact color histogram descriptor represented in the HSV
  color space[3] and encoded using Haar transform.
- **Texture descriptors** The *homogeneous texture descriptor* characterizes the
  regional texture using local spatial frequency statistics extracted by Gabor filter
  banks. The *texture browsing descriptor* represents a perceptual characterization
  of texture in terms of regularity, coarseness, and directionality as a vector. The
  *edge histogram descriptor* represents the local edge distribution of an image as a
  histogram that corresponds to the frequency and directionality of brightness
  changes in the image.
- **Shape descriptors** The *region-based shape descriptor* represents the distribution of all interior and boundary pixels that constitute a shape by decomposing
  the shape into a set of basic functions with various angular and radial frequencies
  using angular radial transformation, a two-dimensional complex transform
  defined on a unit disk in polar coordinates. The *contour-based shape descriptor*
  represents a closed two-dimensional object or region contour in an image or

---

[2]https://www.mathworks.com/products/computer-vision.html

[3]Low-level color descriptors often use nonlinear color spaces, because those are closely related to
the human perception of color. One such color space is HSV, which represents colors with hue–
saturation–value triplets. Another one is HMMD, which holds the hue, the maximum and the
minimum of RGB values, and the difference between the maximum and minimum values. MPEG-
7 supports both of these color spaces, in addition to the widely used RGB, YCbCr, and mono-
chrome color spaces.

video. The *3D shape descriptor* is a representation-invariant description of three-dimensional mesh models, expressing local geometric attributes of 3D surfaces defined in the form of shape indices calculated over a mesh using a function of two principal curvatures.

- **Motion descriptors** The *camera motion descriptor* represents global motion parameters, which characterize a video scene in a particular time by providing professional video camera movements, including moving along the optical axis (dolly forward/backward), horizontal and vertical rotation (panning, tilting), horizontal and vertical transverse movement (tracking, booming), change of the focal length (zooming), and rotation around the optical axis (rolling). The *motion activity descriptor* indicates the intensity and direction of motion, and the spatial and temporal distribution of activities. The *motion trajectory descriptor* represents the displacement of objects over time in the form of spatiotemporal localization with positions relative to a reference point and described as a list of vectors. The *parametric motion descriptor* describes the global motion of video objects using a classic parametric model (translational, scaling, affine, perspective, quadratic).

### 3.1.2 Common Audio Descriptors

The most common audio descriptors are the following:

- **Temporal audio descriptors** The *energy envelope descriptor* represents the root mean square of the mean energy of the audio signal, which is suitable for silence detection. The *zero crossing rate descriptor* represents the number of times the signal amplitude undergoes a change of sign, which is used to differentiate periodic signals and noisy signals, such as to determine whether the audio content is speech or music. The *temporal waveform moments descriptor* represents characteristics of waveform shape, including temporal centroid, width, asymmetry, and flatness. The *amplitude modulation descriptor* describes the tremolo of a sustained sound (in the frequency range 4–8 Hz) or the graininess or roughness of a sound (between 10 and 40 Hz). The *autocorrelation coefficient descriptor* represents the spectral distribution of the audio signal over time, which is suitable for musical instrument recognition.
- **Spectral audio descriptors** The *spectral moments descriptor* corresponds to core spectral shape characteristics, such as spectral centroid, spectral width, spectral asymmetry, and spectral flatness, which are useful for determining sound brightness, music genre, and categorizing music by mood. The *spectral decrease descriptor* describes the average rate of spectral decrease with frequency. The *spectral roll-off descriptor* represents the frequency under which a predefined percentage (usually 85–99%) of the total spectral energy is present, which is suitable for music genre classification. The *spectral flux descriptor* represents the dynamic variation of spectral information computed either as the

normalized correlation between consecutive amplitude spectra or as the derivative of the amplitude spectrum. The *spectral irregularity descriptor* describes the amplitude difference between adjacent harmonics, which is suitable for the precise characterization of the spectrum, such as for describing individual frequency components of a sound. The descriptors of *formants parameters* represent the spectral peaks of the sound spectrum of voice and are suitable for phoneme and vowel identification.

- **Cepstral audio descriptors** Cepstral features are used for speech and speaker recognition and music modeling. The most common cepstral descriptors are the *mel-frequency cepstral coefficient descriptors*, which approximate the psychological sensation of the height of pure sounds, and are calculated using the inverse discrete cosine transform of the energy in predefined frequency bands.
- **Perceptual audio descriptors** The *loudness descriptor* represents the impression of sound intensity. The *sharpness descriptor*, which corresponds to a spectral centroid, is typically estimated using a weighted centroid of specific loudness. The *perceptual spread descriptor* characterizes the timbral width of sounds, and is calculated as the relative difference between the specific loudness and the total loudness.
- **Specific audio descriptors** The *odd-even harmonic energy ratio descriptor* represents the energy proportion carried by odd and even harmonics. The descriptors of *octave band signal intensities* represent the power distribution of the different harmonics of music. The *attack duration descriptor* represents how quickly a sound reaches full volume after it is activated and is used for sound identification. The *harmonic-noise ratio descriptor* represents the ratio between the energy of the harmonic component and the noise component, and enables the estimation of the amount of noise in the sound. The *fundamental frequency descriptor*, also known as the *pitch descriptor*, represents the inverse of the period of a periodic sound.

### 3.1.3  Common Spatiotemporal Feature Descriptors, Feature Aggregates, and Feature Statistics

Local spatiotemporal feature descriptors, aggregates, and statistics, which capture both appearance and motion, have become popular for action recognition in videos. The most common ones are the following:

- *SIFT (Scale-Invariant Feature Transform)* [59] As the name suggests, SIFT is a scale-invariant feature descriptor, which is suitable for object recognition, robotic navigation, 3D modeling, gesture recognition, and video tracking.
- *Cuboid Descriptor* [60] The cuboid descriptor is a spatiotemporal interest point detector, which finds local regions of interest in space and time (cuboids) to be used for behavior recognition.

- *HOG (Histogram of Oriented Gradients)* [61] HOG describes the number of occurrences of gradient orientation in localized portions of images. HOG tends to outperform many other feature descriptors.
- *HOF (Histogram of Optical Flow)* [62] HOG-based appearance descriptors combined with various HOF-based motion descriptors are suitable for detecting humans in videos.
- *MBH (Motion Boundary Histograms)* MBH captures local orientations of motion edges by emulating static image HOG descriptors.
- *SURF (Speeded Up Robust Features)* [63] The SURF feature descriptor is based on the sum of the Haar wavelet response around the point of interest. It can be used to locate and recognize objects and people, reconstruct 3D scenes, track objects, and extract points of interest.

## 3.2 The Discrepancy Between Low-Level Features and High-Level Semantics

The main reason behind the difficulty of multimedia understanding is the huge semantic gap between what computers can interpret via automatically extracted low-level features and what humans can understand based on cognition, knowledge, and experience. While useful, many automatically extracted low-level multimedia features are inadequate for representing multimedia semantics. For example, motion vectors or dominant colors do not provide information about the meaning of the visual content (see Fig. 3.1).

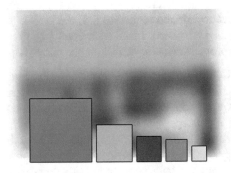

**Fig. 3.1** Many automatically extracted low-level features, such as dominant colors (*left*), are insufficient to interpret the visual content (*right*)

In contrast, high-level descriptors are suitable for multimedia concept mapping, yet they often rely on human knowledge, experience, and judgment. Manual video concept tagging is very time-consuming, might be biased, too generic, or inappropriate, which has led to the introduction of collaborative semantic video annotation, wherein multiple users annotate the same resources and improve each other's annotations [64]. Manual annotations can be curated using natural language processing to eliminate duplicates and typos, and filter out incorrectly mapped concepts. The integrity of manual annotations captured as structured data can be checked automatically using LOD definitions. Research results for high-level concept mapping in constrained videos, such as medical videos [65] and sport videos [66], are already promising; however, concept mapping in unconstrained videos is still a challenge [67].

The major approaches undertaken to narrow the semantic gap between low-level features and high-level multimedia semantics encode human knowledge in knowledge-based information systems using logical formalisms based on one or more of the following:

- *Mid-level feature extraction* Mid-level features, such as object class hierarchy, visual patterns, segmented image patches, tracked objects, bags of features, spatial pyramids, and named entities extracted from subtitles, represent perceptual intuitions as well as higher-level semantics derived from signal-level saliency. Mid-level features are suitable for constructing expressive semantic features for general visual content understanding and classification.
- *Information fusion* The combination of multiple features from different modalities, such as video, audio, and subtitles, can provide rich semantics and outperform pretrained classifiers based on ad hoc low-level features. Two main types of information fusion exist:
  - *Early fusion* Features of different modalities are combined and used as an input for machine learning to extract combined semantics [68, 69].
  - *Late fusion* Extracted mid-level audio and video semantics are combined for a higher level of abstraction and improved semantic extraction accuracy [70].
- *Calculating semantic relatedness* Context retrieved from the relationships between depicted concepts is very useful for the semantic enrichment of videos. For example, corpora of concepts frequently depicted together can be used to set likelihood values for the correctness of concept mapping.
- *Utilizing very expressive ontologies* Many multimedia ontologies cover a small knowledge domain only, exploit a very small subset of the constructors available in the underlying knowledge representation language, or provide low-level features only. The precise annotation of video scenes requires very expressive high-level descriptors for the represented domain, complemented by complex rules.
- *Utilizing SWRL rules* Rule-based formalisms are suitable for annotating video events and provide very rich semantics about the knowledge domain related to

the video content. For example, by formally defining the competition rules of soccer, the number of players, the layout of the pitch, and match events, it is possible to automatically generate subtitles for soccer videos [71].

## 3.3 Semantic Enrichment of Multimedia Resources

Some de facto standard and standard metadata specifications are designed for a particular media content type and others for multiple media content types. The most common general-purpose metadata specifications are the following:

- *Dublin Core* is one of the most versatile general-purpose metadata standards (IETF RFC 5013, ISO 15836-2009, NISO Z39.85). It can describe general metadata, such as the abstract, creator, date, publisher, title, and language of web resources (websites, images, videos), physical resources (books, CDs, DVDs), and objects such as artworks. Dublin Core is widely deployed as semistructured metadata embedded to images and multimedia files, traditional website markup,[4] and structured data on the Semantic Web.
- *Creative Commons* is the de facto standard for general-purpose licensing metadata, which can also be used for image, audio, and video licensing, although there are more specific standards that are dedicated to multimedia licensing, such as MPEG-21, as you will see later.

*Schema.org* is the largest collection of structured data schemas on the Web. It was launched by Google, Yahoo!, and Bing in 2011. Schema.org contains the machine-readable definition of common concepts, enabling the annotation of actions, creative works, events, services, medical concepts, organizations, persons, places, and products.

Similar to multimedia standards, there are generic and specific OWL ontologies for multimedia resource descriptions. Some of the generic multimedia ontologies are the *Multimedia Metadata Ontology (M3O)*,[5] which was designed to integrate the core aspects of multimedia metadata [72], and W3C's *Ontology for Media Resources*,[6] which provides a core vocabulary with standards alignment to be used in online media resource descriptions.

The standards and ontologies dedicated to a specific media content type are discussed in the following sections.

---

[4]In XHTML 1.x. Prior to HTML5, the namespace mechanism required for such annotations was not supported by HTML, only XHTML.

[5]http://m3o.semantic-multimedia.org/ontology/2009/09/16/annotation.owl

[6]https://www.w3.org/TR/mediaont-10/

### 3.3.1   Semantic Enrichment of Images

While still images are not multimedia resources, they are frequently used in combination with other content forms, such as text, audio, animation, video, and interactive content, to create multimedia content. The core standards used for annotating still images, image sequences, and individual video frames are summarized in the following sections.

#### 3.3.1.1   Core Image Metadata Standards

Most digital cameras can save exposure and camera setting metadata in the *Exchangeable Image File Format (EXIF)*. Examples of such metadata include shutter speed, aperture, metering mode, exposure compensation, flash use, camera make and model, and lens. Another metadata format, Adobe's *XMP*, specifies a data model, an XML serialization, and formal schema definitions for metadata management. The XMP data model is an RDF-based data model, and the XMP schemas define sets of metadata property definitions, covering basic rights and media management, while utilizing Dublin Core for general multimedia metadata [73]. When edited in compliant image editing applications, such as Adobe Photoshop, XMP and EXIF metadata are serialized as RDF/XML and written directly to compatible binary image files, such as JPEG [74] (see Fig. 3.2).

**Fig. 3.2** Metadata edited in Adobe Photoshop (*foreground*) is saved directly to the binary JPG image file (*background*)

*DIG35* is another image metadata specification, which provides not only general image metadata, but also properties for content description, file history, and image licensing.

#### 3.3.1.2  Image Vocabularies and Ontologies

The US Library of Congress, the largest library in the world, developed a thesaurus in XML with more than 7000 terms for indexing visual materials (photographs, prints, design drawings, etc.) by subject, and 650 terms for indexing visual materials by genre or format, called the *Thesaurus for Graphic Materials (TGM)*.[7,8]

From the aforementioned image metadata standards, the vocabulary of EXIF and DIG35 has been mapped to Semantic Web standards: the EXIF vocabulary has been mapped to an RDFS ontology, with domains and ranges defined for the properties,[9] and the DIG35 vocabulary has been mapped to an OWL Full ontology.[10]

### *3.3.2  Structured 3D Model Annotation*

The precise reconstruction of real-world 3D objects requires shape measurements and spectrophotometric property acquisition, typically performed using 3D laser scanners, RGB-D depth cameras, Kinect depth sensors, structured light devices, photogrammetry, and photomodeling [75]. Not only the dimension and shape have to be captured, but also the textures, diffuse reflection, transmission spectra, transparency, reflectivity, opalescence, glazes, varnishes, enamels, and so on. Many of these properties can be represented by low-level descriptors [76], which can be used for training machine learning [77], but these are limited when it comes to machine processability and do not provide the meaning of the corresponding 3D model or scene. These issues can be addressed using OWL ontologies and LOD interlinking.

3D models derived from 3D scanning or 3D modeling can be efficiently annotated using structured data. However, many of the 3D modeling file formats are not structured, and as such, cannot be used directly for structured 3D annotations. For example, the *Virtual Reality Modeling Language (VRML)* and the *Polygon File Format (PLY)* describe 3D models in plain text files (unstructured data). The successor of VRML, *Extensible 3D (X3D)*, is now the industry standard for 3D models (ISO/IEC 19775, 19776, and 19777), and can be used for representing interactive 3D computer graphics in web browsers without proprietary plugins.

---

[7]https://www.loc.gov/rr/print/tgm1/tgm1.xml

[8]https://www.loc.gov/rr/print/tgm2/tgm2.xml

[9]http://www.kanzaki.com/ns/exif

[10]http://users.datasciencelab.ugent.be/gmartens/Ontologies/DIG35/v0.2/DIG35.owl

Written in XML Schema, the original X3D vocabulary was semistructured. Over the years, part of this vocabulary was mapped to OWL (e.g., *OntologyX3D* [78], the Petit mapping of X3D [79]), and more recently the entire vocabulary to OWL 2 with extensions (*3D Modeling Ontology*[11] [80]).

The spatial fragmentation of 3D models is less researched. Although the MPEG-21 standard specifies URI-based fragment identifiers for annotating 3D spatial regions, they are only available for MPEG-21-compliant time-based multimedia files[12] stored in a format based on the ISO Base Media File Format (MPEG-4 Part 12, ISO/IEC 14496-12). Furthermore, the MPEG-21 standard does not provide implementation examples for arbitrary 3D objects and does not support viewpoint data.

### 3.3.3   Semantic Enrichment of Audio and Video

OWL ontologies are proven to be efficient in the representation of audio events [81], mood, emotions, and concepts associated with music [82], as well as contextual information and video semantics, which are suitable for the advanced interpretation of video scenes [83]. Domain ontologies can interlink machine-readable definitions of real-world concepts depicted in videos with LOD datasets and separate domain knowledge from administrative and technical metadata. The resulting structured video metadata can be efficiently modeled in RDF and implemented in the website markup, as a separate file, or in LOD datasets [84]. Ontologies provide semantic enrichment not only for videos of video repositories and video objects embedded in web pages, but also for hypervideo applications and video streaming portals, such as YouTube [85]. However, many multimedia ontologies provide very generic concept definitions only, such as title and description, and often focus on movie character-istics, such as the crew members who participated in making the film and the film studio where the film was made. While such properties are suitable for lightweight video annotations of Hollywood movies, they are inefficient for annotating the characteristics of music clips, sports videos, commercials, reports distributed through webcasts and telecasts, educational videos, and medical videos, and do not cover the technical features of arbitrary video files, such as video mode, chroma subsampling, and passband modulation.

#### 3.3.3.1   Core Audio and Video Metadata Standards

In parallel with the tremendously increasing number of multimedia contents on the Web, many technical specifications and standards have been introduced to store

---

[11]http://3dontology.org

[12]http://standards.iso.org/ittf/PubliclyAvailableStandards/c050551_ISOIEC_21009_2005_Amd_2008_Annex_D.pdf

technical details and describe the content of multimedia resources. Beyond the unstructured proprietary tags embedded in multimedia files, multimedia metadata specifications have been standardized over the years for generic multimedia metadata and the spatial, temporal, and spatiotemporal annotation of videos. *ID3* is the de facto metadata standard for MP3 files, which provides useful information, such as title, artist, year, and genre,[13] to be displayed by compatible software and hardware (media player software, portable MP3 players, car stereo, etc.) during audio playback. *MPEG-7* is an ISO standard (ISO/IEC 15938),[14] which provides XML metadata to be attached to the timecode of MPEG-1, MPEG-2, and MPEG-4 contents (e.g., synchronized lyrics for music videos). *MPEG-21* (ISO/IEC 21000)[15] provides machine-readable licensing information for MPEG contents in XML. *NewsML* (IPTC NewsML-G2)[16] is a media-independent, structural framework for multimedia news. *TV-Anytime* (IETF RFC 4078,[17] ETSI TS 102 822)[18] was designed for the controlled delivery of personalized multimedia content to consumer platforms.

### 3.3.3.2 Vocabularies and Ontologies

The vocabularies of the aforementioned audio and video metadata specifications have originally been created in XML or XML Schema (XSD),[19,20,21] which makes these vocabularies machine-readable, but not machine-interpretable. As a result, the previous standards in their original form of release are inefficient for automated content access, sharing, and reuse. These limitations can be addressed using Semantic Web standards, such as RDF, RDFS, and OWL, rather than XML, for multimedia annotations. Consequently, ID3, Dublin Core, TV-Anytime, and MPEG-7 have since been mapped to Semantic Web standards,[22,23,24,25] and several OWL ontologies have been created with or without standards alignment [86].

In the 2000s, several attempts have been made to address the semantic limitations of MPEG video content descriptors by mapping the MPEG-7 concepts defined in XSD to OWL [87]. Since the introduction of MPEG-7, the research community

[13]http://id3.org

[14]https://www.iso.org/standard/34230.html

[15]https://www.iso.org/standard/35367.html

[16]https://iptc.org/standards/newsml-g2/

[17]https://tools.ietf.org/html/rfc4078

[18]http://www.etsi.org/technologies-clusters/technologies/broadcast/tv-anytime

[19]https://www.iso.org/standard/34230.html

[20]http://purl.org/NET/mco-core, http://purl.org/NET/mco-ipre

[21]https://portal.etsi.org/webapp/workprogram/Report_WorkItem.asp?WKI_ID=39864

[22]http://ansgarscherp.net/ontology/m3o.semantic-multimedia.org/mappings/id3/id3.owl

[23]http://dublincore.org/2012/06/14/dcterms.rdf

[24]http://rhizomik.net/ontologies/2005/03/TVAnytimeContent.owl

[25]http://mpeg7.org/mpeg7.ttl

has focused on the standard because of its capability to describe low-level features; despite this, the corresponding descriptors are often insufficient for multimedia content description. For example, if the automatically extracted dominant colors of an image are red, black, and yellow, it cannot be inferred that the image depicts a sunset [88]. Moreover, MPEG-7 provides machine-readable syntactic metadata, but not machine-interpretable semantics. This limitation was addressed by OWL ontologies based on the MPEG-7 standard. Hunter was the first to model the core parts of MPEG-7 in OWL Full [89]. This approach utilized RDF to formalize the core parts of MPEG-7, which was complemented by DAML+OIL constructs to further detail the semantics. Tsinaraki et al. extended this ontology by including the full Multimedia Description Scheme (MDS) of MPEG-7 [90]. Furthermore, Isaac and Troncy proposed a core audiovisual ontology based on MPEG-7, ProgramGuideML, and TV-Anytime [91]. The first complete MPEG-7 ontology, *Rhizomik* (MPEG-7Ontos), was generated using an XSD to OWL mapping in combination with a transparent mapping from XML to RDF [92]. Blöhdorn et al. proposed the *Visual Descriptor Ontology (VDO)* for image and video analysis covering the visual components of MPEG-7 in OWL DL [93]. The *Multimedia Structure Ontology (MSO)* defined basic multimedia entities from the MPEG-7 MDS and mutual relations such as decomposition [94]. The two OWL DL ontologies of Dasiopoulou et al., the *Multimedia Content Ontology (MCO)* and the *Multimedia Descriptors Ontology (MDO)*, cover the MPEG-7 MDS structural descriptors, and the visual and audio parts of MPEG-7 [95]. Oberle et al. created an ontological framework to formally model the MPEG-7 descriptors and export them into OWL and RDF, based on the structural, localization, media, and low-level descriptors [96].

Furthermore, music ontologies are utilized not only in indexing, but also by applications to provide customized services, such as by automatically finding similar music artists (e.g., DBTune's *Last.fm Artist Similarity RDF Service*)[26] or aggregating structured data according to the music collection of a user (e.g., gnarql).[27] The *Audio Features Ontology*[28] represents automatically extracted low-level audio features and utilizes specific terms defined in domain ontologies. The *Tonality Ontology*[29] is suitable for defining high-level and low-level descriptors for tonal content in RDF. The *Keys Ontology*[30] is designed for describing keys in musical pieces (e.g., C major, A♭ minor). The *Chord Ontology*[31] is an ontology for describing chords and chord sequences of musical pieces in RDF. The *Music Ontology*[32] is suitable for the annotation of music-specific audio features (e.g., artist, composer, conductor, discography, record, remixer, singer, and tempo). Both the Audio Features Ontology and the Music Ontology rely on the *Event Ontology*[33]

---

[26]https://github.com/motools/LFM-Artist-Similarity-RDF-Service

[27]https://github.com/motools/gnarql

[28]http://purl.org/ontology/af/

[29]http://purl.org/ontology/tonality/

[30]http://purl.org/NET/c4dm/keys.owl

[31]http://purl.org/ontology/chord/

[32]http://purl.org/ontology/mo/

[33]http://purl.org/NET/c4dm/event.owl#

and the *Timeline Ontology*.[34] The *MusicBrainz Schema*[35] provides structured annotations for music artists, events, and recordings. Kanzaki's *Music Vocabulary*[36] can be used to describe classical music and performances, including musical works, events, instruments, performers, and their properties (e.g., `Concert`, `Conductor`, `Orchestra`, `Symphony`, `Composer`), and to distinguish musical works (e.g., `Opera`, `String_Quartette`) from performance events (e.g., `Opera_Event`) and performers (e.g., `StringQuartetEnsemble`). The *Classical Music Navigator Ontology*[37] describes influences between classical composers. The *Temperament Ontology*[38] can be used to describe musical instrument tuning systems and their particularities, as well as to characterize a temperament used when tuning an instrument for a particular performance or recording. The *Multitrack Ontology*[39] is an ontology for describing concepts in multitrack media production. The *Audio Effects Ontology*[40] *(AUFX-O)* provides concepts and properties for audio effects used in professional music production.

*Schema.org* provides de facto standard structured annotations for a variety of knowledge domains, including the multimedia domain. For example, a single song can be described using `schema:MusicRecording`, a collection of music tracks with `schema:MusicAlbum`, a collection of music tracks in playlist form with `schema:MusicPlaylist`, and the technical metadata of audio resources with properties such as `schema:bitrate`, `schema:encodingFormat`, and `schema:duration`. Similarly, generic video metadata can be provided for video objects using `schema:video` and `schema:VideoObject`, and movies, film series, seasons, and episodes of series can be described using `schema:Movie`, `schema:MovieSeries`, `schema:CreativeWorkSeason`, and `schema:Episode`, respectively. All of these annotations can be serialized in RDFa, JSON-LD, and HTML5 Microdata, which are indexed by all the major search engines, including Google, Yahoo!, and Bing.

The genre of music and movies can be defined using `schema:genre`. Music genres can also be declared using the aforementioned Music Ontology. A comprehensive controlled vocabulary for genres is the *EBU Genre Vocabulary*, which is a superset of the genre classification concepts of MPEG-7 and TV-Anytime, although it is defined in XML.[41] For applications such as age rating of movies, movie cataloging, and TV program description, the semantic description of video scenes can be improved with cues about negative, neutral, and positive emotions; emotional responses; affection; and sentiment by utilizing emotion ontologies, such as *STIMONT* [97].

---

[34]http://purl.org/NET/c4dm/timeline.owl#

[35]http://musicbrainz.org/doc/MusicBrainz_Database/Schema

[36]http://www.kanzaki.com/ns/music

[37]http://purl.org/ontology/classicalmusicnav

[38]http://purl.org/ontology/temperament/

[39]http://purl.org/ontology/studio/multitrack

[40]https://w3id.org/aufx/ontology/1.0

[41]https://www.ebu.ch/metadata/cs/ebu_ContentGenreCS.xml

### 3.3.3.3   Issues and Limitations of the XSD-OWL Mappings of MPEG-7

While several attempts have been made to map MPEG-7 terms to OWL, all these mappings inherited the following issues from MPEG-7:

- Structural complexity: the MPEG-7 XML schemas define 1182 elements, 417 attributes, and 377 complex types
- Lack of transparency: the standard is not freely available and its documentation is poor, making the specification difficult to implement
- Vocabulary limited to low-level descriptors: MPEG-7-based ontologies rely on upper ontologies for subject matter metadata and domain ontologies for high-level semantics
- Conceptual ambiguity: semantically equivalent descriptors representing the same information are defined by multiple classes and entities,[42] as, for example, mpeg7:Keyword and mpeg7:StructuredAnnotation
- Semantic interoperability issues: lack of explicit axiomatization of the intended semantics
- Syntactic interoperability issues: unaddressed syntactic variability

Suárez-Figueroa et al. summarized the limitations of MPEG-7-based and domain ontologies as follows [98]. Hunter's OWL ontology was limited to MPEG-7 upper MDS and did not define inverse relationships. Tsinakari's MPEG-7-based ontology used an inconsistent naming convention. The Multimedia Structure Ontology had conceptual ambiguity issues and did not define disjointness.[43] Rhizomik, while the first ontology with full MPEG-7 coverage, does not define property domains and ranges, uses an inconsistent naming convention, and some of its elements share the same URI.

Arndt et al. introduced an ontology for multimedia annotation, the *Core Ontology for Multimedia (COMM)*, by adding formal semantics to selected parts of MPEG-7 [99]. Although COMM is one of the very few multimedia ontologies that exploited all the mathematical constructors available in the underlying formal language at the time of its release, it does not define the domain and range for all the properties, nor all the relationships, such as disjointness.

To overcome the previous limitations, domain ontologies not or not entirely based on MPEG-7 have also been proposed in the literature for movie annotation, scene representation, and video concept detection. The ontology introduced by

---

[42]This is the result of incorrect modeling; not to be confused with *punning*, i.e., denoting different entity types (an individual and a class) in an ontology with the same identifier (used for metamodeling).

[43]Unfortunately, not exploiting expressive mathematical constructors available in the underlying representation language has always been a common practice in multimedia ontology engineering. While low expressivity might be used purposefully to achieve low computational complexity, it also limits the tasks that can be performed on ontologies.

Zha et al. focuses exclusively on video concept detection [100], while the *Linked Movie Database*[44] was designed specially for expressing movie features (common concepts and properties of Hollywood films, such as actor, director, etc.), leaving many video characteristics uncovered.

Similar to other multimedia ontologies, the Linked Movie Database ontology is limited by ambiguous, imprecise, and overlapping terms. For example, the property value examples for `movie:film_format`, such as `Silent film`, `Animation`, `MGM Camera 65`, `VistaVision`, `Circle-Vision 360°`, `IMAX`, `CinemaScope`, `3-D film`, `High-definition video`, `Anamorphic widescreen`, reveal that `movie:film_format`[45] is not precise, because it mixes a variety of video file and movie features, namely:

- MGM Camera 65 and CinemaScope refer to the recording technology
- Circle-Vision 360° refers to the playback technology
- IMAX refers to both the recording and the playback technology
- 3D and silent represent film characteristics
- High-definition refers to video quality

Moreover, this lightweight domain ontology does not provide a comprehensive coverage even for the very domain it was created, and LMB properties such as `film_regional_release_date`, `film_genre`, and `film_screening_venue` cannot be used for videos other than films.[46] The *Video Ontology (VidOnt)* is a core reference ontology for video modeling with MPEG-7 alignment, addressing some of the aforementioned issues via formally grounded conceptual modeling and high expressivity.

### 3.3.3.4   Automated Semantic Video Annotation

While the powerful Semantic Web standards suitable for the formal definition of multimedia semantics can contribute to better multimedia representation and understanding, manual annotation is often infeasible, and automated multimedia annotation remains a challenge. Video ontologies and rulesets can describe even complex video scenes; however, the automatically extractable visual features of objects in video shots represented using low-level descriptors (such as that of MPEG-7) do not correspond to high-level concepts, but only to "semi-concept" values that might be used to infer high-level concepts. Shot-level inference rules applied to ontologies can sometimes automatically extract high-level concepts, and semantic scene description can be generated based on the frequency and similarity of high-level concepts from groups of semantically similar concepts [101].

---

[44]http://www.linkedmdb.org

[45]http://data.linkedmdb.org/directory/film_format

[46]Moreover, the naming convention of these properties does not comply with best practices, which will be demonstrated in the next chapter.

Using information fusion, high-level concepts retrieved from subtitles and audio can be combined with the concepts identified in the video to obtain complementary information. Neither the early fusion methods, which merge multimodal features into a longer feature vector before feeding them to the classifiers, nor the late fusion methods, which combine the detection outputs after generating multiple unimodal classifiers, are perfect. While early fusion can model the correlation between different feature components by combining them into long feature vectors, overly heterogeneous feature vectors with skewed length distribution and numerical scales are problematic. Late fusion is better from this point of view, because the features from the different modalities do not interact with each other before the final fusion stage. Also, late fusion enables specific detection techniques for the different feature types and is less computation-intensive than early fusion, which make late fusion the more popular of the two fusion methods. The confidence scores and features generated from different modalities are usually normalized before fusion with rank, range, logistic, or Gaussian normalization. The final detection results are obtained by merging the normalized confidence values using one or more input types, including the combination of multiple detection models, the combination of the detection models of the same class with different underlying features, or the combination of the models with the same underlying features and different parameters. The combination methods include predefined combination rules based on domain knowledge and a variety of machine learning methods. The analysis and integration of high-level concepts yields to a (rough) video summary.

## 3.4  Summary

This chapter introduced the reader to the semantic gap, the discrepancy between the automatically extractable low-level features and the high-level semantics that define the meaning of audio waveforms and scenes of 3D models, images, and videos. The most common visual and audio descriptors have been described, along with common spatiotemporal feature descriptors, feature aggregates, and feature statistics. The semantic enrichment of multimedia resources has been demonstrated, and the core metadata standards and ontologies listed.

# Chapter 4
# Description Logics: Formal Foundation for Web Ontology Engineering

Logic-based knowledge representations implement mathematical logic, a subfield of mathematics dealing with formal expressions, reasoning, and formal proof. *Description logics (DL)* constitute a family of formal knowledge representation languages used for ontology grounding, which enables automated reasoning over RDF statements. Many description logics are more expressive than propositional logic, which deals with declarative propositions and does not use quantifiers, and more efficient in decision problems than first-order predicate logic, which uses predicates and quantified variables over nonlogical objects. Beyond decidability, a crucial design principle in description logics is to establish favorable trade-offs between expressivity and scalability, and when needed, maximize expressivity. This is one of the reasons behind the variety of description logics, because the best balance between expressivity of the language and the complexity of reasoning depends on the intended application. The expressivity of DL-based ontologies is determined by the mathematical constructors available in the underlying description logic. Beyond the general-purpose description logics, multimedia descriptions may utilize spatial and temporal description logics as well.

## 4.1 Description Logics

Knowledge engineers, who simulate human decision-making and high-level cognitive tasks in computer systems, use logic-based approaches to formally represent, and enable the automated processing of, human knowledge. *Description logics (DL)* are the successors of early AI approaches for knowledge representation, such as semantic networks and frames. In contrast to their predecessors, however, description logics are equipped with formal, logic-based semantics that, among others, provide formal grounding for ontology languages, such as OWL. Description logics are related to other logics, such as first-order logic (FOL) and modal logic (ML). One of the main benefits of description logics is that their computational properties

© Springer International Publishing AG 2017
L.F. Sikos, *Description Logics in Multimedia Reasoning*,
DOI 10.1007/978-3-319-54066-5_4

are well understood, among which worst-case complexity is particularly important.[1] Description logics provide logical formalisms for practical decision procedures used to solve common problems, such as satisfiability, subsumption, and query answering, and are implemented in highly optimized automated systems. Ontology-based scene interpretation relies on the formal description of scenes, for which description logics are suitable, such as by mapping atomic concepts to geometric primitives [102].

Description logic axioms usually capture only partial knowledge about the knowledge domain the ontology describes, and there may be multiple states of the world that are consistent with an ontology. DL-based ontologies do not specify a particular interpretation based on a default assumption or a state of the world; instead they consider all possible cases in which the axioms are satisfied. This characteristic makes it possible to handle incomplete information, such as statements that have not been explicitly made yet, by keeping unspecified information open, which is called the *open world assumption (OWA)*. The truth value of a description logic statement may be true irrespective of whether or not it is known to be true, and irrespective of whether we believe that it is true or not. Consequently, the open world assumption not only limits semantic reasoners in terms of deduction potential , but also prevents them from making false deductions: from the absence of a statement alone a deductive reasoner will not infer that the statement is false. Since there are scenarios where the *closed world assumption (CWA)* is desired, however, proposals have been made for modeling CWA systems with OWL, such as by using DL-safe variables[2] [103], axioms with CWA seen as special cases of integrity constraints [104], and combining OWL with closed world rule-based languages [105].

### 4.1.1   Nomenclature

While FOL, DL, and OWL are related, there are terminological differences between them. Most DL concepts correspond to FOL unary predicates, DL roles to FOL binary predicates, DL individuals to FOL constants, DL concept expressions to FOL formulae with one free variable, role expressions to FOL formulae with two free variables, a DL concept corresponds to an OWL class, and a DL role to an OWL property, as shown in Table 4.1.

---

[1]See Chap. 6 for more details.

[2]Each variable in a DL-atom must also occur in a non-DL-atom (restriction to explicitly introduced individuals).

**Table 4.1** Terminology comparison of FOL, DL, and OWL

| FOL | DL | OWL |
|---|---|---|
| Unary predicate | Atomic concept (concept name) | Class name |
| Formula with one free variable | Concept | Class |
| Binary predicate | Atomic role (role name) | Property name |
| Formula with two free variables | Role | Property |
| Constant | Individual | Individual |
| Sentence | Axiom | Axiom |
| Signature | Vocabulary or signature | Controlled vocabulary |
| Theory | Knowledge base | Ontology |

## *4.1.2 Annotation and Naming Conventions*

The most common notational conventions of description logics are as follows. In definitions, the atomic concepts are usually denoted by $A$ and $B$, concept descriptions by $C$ and $D$, and roles by $R$ and $S$. Individuals are represented by $a$ and $b$. Functional roles (features/attributes) are denoted by $f$ and $g$. In cardinality restrictions, the nonnegative integers are typically denoted by $n$ and $m$. Sets are denoted by bold font.

In concrete examples, concept names should be written in PascalCase, i.e., the first letter of the identifier and the first letter of each subsequent concatenated word are capitalized (e.g., BroadcastService, MovieClip). Role names are typically written in camelCase, i.e., the first letter of the identifier is lowercase and the first letter of each subsequent concatenated word is capitalized (e.g., directedBy, portrayedBy).[3] Individual names are written in ALL CAPS (e.g., ZAMBEZIA, JACKIECHAN).

The de facto standard font for description logic names is the cursive font *CMSY10*. The letters in description logic names indicate the set of available mathematical constructors by appending the corresponding letter (see Table 4.2), except when using a letter representing a superset of another letter, in which case the letter of the subset is omitted.[4]

For example, $\mathcal{SHIN}$ is a superset of $\mathcal{AL}$, $\mathcal{C}$, $\mathcal{H}$, $\mathcal{I}$, and $\mathcal{N}$, and includes all the mathematical constructors of $\mathcal{AL}$ and $\mathcal{C}$, as well as transitive roles (i.e., $\mathcal{S}$), role

---

[3]The description logic concepts and roles do not follow the general capitalization rules of English grammar, but purposefully capitalize each word to make them easier to read. For example, according to grammatical rules, prepositions fewer than five letters (e.g., on, at, to, from, by) are not capitalized unless they are the first words in a title, while they are capitalized in description logic concept and role names, as shown above.

[4]All description logic names and annotations in this book follow the standard DL nomenclature to avoid confusion. Authors of special description logics often use the CMSY10 font inconsistently, where some letters of a description logic name do not correspond to mathematical constructors. The CMSY10 font is used throughout this book according to the standard DL nomenclature, and DL names were rewritten wherever needed.

**Table 4.2** Common constructors in description logics

| Symbol | Constructors | Example |
|---|---|---|
| $\mathcal{AL}$ | Atomic negation, concept intersection, universal restrictions, limited existential quantification | MovieCharacter ⊓ Star |
| $\mathcal{C}$ | Complement (complex concept negation) | ¬(LiveAction ⊔ ComputerAnimation) |
| $\mathcal{S}$ | $\mathcal{ALC}$ with transitive roles | partOf ∘ partOf ⊑ partOf |
| $\mathcal{E}$ | Full existential quantification | ∃hasDiscRelease.Movie |
| $\mathcal{F}$ | Role functionality, a special case of uniqueness quantification | ⊤ ⊑ ⩽1officialWebsite.⊤ |
| $\mathcal{U}$ | Concept union | WebcastChannel ⊓ TelevisionChannel |
| $\mathcal{H}$ | Role hierarchy/subproperties | remakeOf ⊑ basedOn |
| $\mathcal{R}$ | Complex role inclusion, reflexivity and irreflexivity, role disjointness | hasParent ∘ hasBrother ⊑ hasUncle |
| $\mathcal{O}$ | Enumerated concept individuals (nominals) | AnimationStudios ≡ { DREAMWORKS, WALTDISNEYANIMATIONSTUDIOS, PIXAR } |
| $\mathcal{I}$ | Inverse roles | directedBy ≈ directorOf⁻ |
| $\mathcal{N}$ | Unqualified cardinality restrictions | TVSeries ≡ ⩾2hasEpisode.Episode |
| $\mathcal{Q}$ | Qualified cardinality restrictions | Actor ⊑= 1hasBirthplace.⊤ |
| $\mathcal{V}$ | Variable nominal classes (nominal schemas) | ∃hasParent.{$z$} ⊓ ∃hasParent.∃married.{$z$} ⊑ $C$ |
| $^{(\mathcal{D})}$ | Datatypes, datatype properties, data values | ⊤⊑∀⩽1maxone.float<br>⊤⊑∀ligthIntensity.maxone<br>lightIntensity(CGIMODEL, 0.75) |

hierarchies ($\mathcal{H}$), inverse roles ($\mathcal{I}$), and the constructors of $\mathcal{R}$. $\mathcal{Q}$ is a superset of $\mathcal{N}$, which is a superset of $\mathcal{F}$, thus $\mathcal{F}$, $\mathcal{N}$, and $\mathcal{Q}$ are never used together in any description logic name. Common description logic names are constructed using the following naming scheme:

$$(([\mathcal{AL}]|(([\mathcal{U}]|[\mathcal{E}])/[\mathcal{C}])))/[\mathcal{S}]|[\mathcal{H}]|[\mathcal{R}]|[\mathcal{O}]|[\mathcal{I}]|([\mathcal{F}]/[\mathcal{N}]/[\mathcal{Q}])|[\mathcal{V}]^{\{\mathcal{D}\}}$$

In this scheme, the round brackets delimit those sets of constructors that are disjoint with other sets of constructors, the square brackets indicate a letter (or two letters in case of $\mathcal{AL}$) that can appear on its own, and the curly brackets indicate an optional superscript.[5] The | character represents *and/or* relation, while the / character represents *or* relation.

---

[5] $^{(\mathcal{D})}$ is the only standard superscript in description logic names that corresponds to a set of constructors from Table 4.2. Other superscripts that add or remove constructors from the sets indicated by the DL name letters include $^+$, $^{++}$, and $^-$. For example, $\mathcal{EL}^+$ extends $\mathcal{EL}$ with transitive closure of roles, $\mathcal{SHOIN}^+$ adds concept existence axioms to $\mathcal{SHOIN}$, and $\mathcal{SR}^+\mathcal{OIQ}$ is the extension of $\mathcal{SROIQ}$ with complex role chains and unions. Special description logics might use further characters that do not correspond to sets of constructors listed in Table 4.2, such as $^{(G)}$, $\star$, etc., some of which are occasionally written as standard characters rather than scripts, as you will see later.

Consequently, the name of the description logic $\mathcal{SROIQ}^{(\mathcal{D})}$, for example, does not contain the letters $\mathcal{A}$, $\mathcal{L}$, $\mathcal{C}$, $\mathcal{H}$, or $\mathcal{N}$, even though it is a superset of the constructors available in $\mathcal{SHOIN}^{(\mathcal{D})}$, $\mathcal{ALCHIQ}^{(\mathcal{D})}$, etc. Also, some description logic constructors are overlapping, as, for example, union and full existential quantification can be expressed using negation, therefore $\mathcal{U}$ and $\mathcal{E}$ are never used together in description logic names, and $\mathcal{C}$ is used instead.

This naming scheme can be demonstrated by analyzing the name of some common description logics. By extending $\mathcal{ALC}$ and transitivity roles (i.e., $\mathcal{S}$) with role hierarchies ($\mathcal{H}$), inverse roles ($\mathcal{I}$), functional properties ($\mathcal{F}$), and datatypes ($\mathcal{D}$), we get the $\mathcal{SHIF}^{(\mathcal{D})}$ description logic, which roughly corresponds to OWL Lite. By adding nominals ($\mathcal{O}$) and cardinality restrictions ($\mathcal{N}$) to $\mathcal{SHIF}^{(\mathcal{D})}$, we obtain $\mathcal{SHOIN}^{(\mathcal{D})}$, the description logic underlying OWL DL. After industrial implementations highlighted several key features missing from $\mathcal{SHOIN}^{(\mathcal{D})}$ to model complex knowledge domains [106], it has been extended with complex role inclusion axioms, reflexive and irreflexive roles, asymmetric roles, disjoint roles, the universal role, self-constructs, negated role assertions, and qualified number restrictions. The result is the very expressive yet decidable $\mathcal{SROIQ}^{(\mathcal{D})}$ description logic [107], which largely corresponds to OWL 2 DL [108]. The extension of $\mathcal{SROIQ}$ with nominal schemas is called $\mathcal{SROIQV}$.

### 4.1.3   Interpretation

The meaning of description logic concepts and roles is defined by their *model-theoretic semantics*, which are based on *interpretations*. These interpretations consist of a domain of discourse, $\Delta$, which is divided into two disjoint sets (an *object domain* $\Delta^{\mathcal{I}}$, also known as an *abstract domain*, and a *datatype domain* $\Delta_{\mathcal{D}}$ of a concrete domain $\mathcal{D}$) and an interpretation function $\cdot^{\mathcal{I}}$. An object domain covers specific abstract objects (individuals) and classes of abstract objects (concepts) as well as abstract roles. A concrete domain $\mathcal{D}$ is a pair $(\Delta_{\mathcal{D}}, \varphi_{\mathcal{D}})$, where $\Delta_{\mathcal{D}}$ is a set called the datatype domain and $\varphi_{\mathcal{D}}$ is a set of concrete predicates. Each predicate name $P$ from $\varphi_{\mathcal{D}}$ is associated with an arity $n$ and an $n$-ary predicate $P^{\mathcal{D}} \subseteq \Delta_{\mathcal{D}}^n$. Concrete domains integrate description logics with concrete sets, such as natural numbers ($\mathbb{N}$), integers ($\mathbb{Z}$), real numbers ($\mathbb{R}$), complex numbers ($\mathbb{C}$), strings, Boolean values, and date and time values, along with concrete roles defined on these sets, including numerical comparisons (e.g., $\leqslant$), string comparisons (e.g., isPrefixOf), and comparisons with constants (e.g., $\leqslant 5$). Mapping objects of the abstract domain to values of the concrete domain via partial functions enables the

modeling of concrete properties of abstract objects, for example, the running time
and release date of movies, and the comparison of these concrete properties.

A Tarski-style DL interpretation $\mathcal{I} = (\Delta^{\mathcal{I}}, \cdot^{\mathcal{I}})$ ("possible world")[6] utilizes a
concept interpretation function $\cdot^{\mathcal{I}_C}$ to map concepts into subsets of the object domain,
a role interpretation function $\cdot^{\mathcal{I}_R}$ to map object roles into subsets of $\Delta^{\mathcal{I}} \times \Delta^{\mathcal{I}}$ and
datatype roles into subsets of $\Delta^{\mathcal{I}} \times \Delta_{\mathcal{D}}$, and an individual interpretation function $\cdot^{\mathcal{I}_I}$ to
map individuals into elements of the object domain (see Fig. 4.1).[7]

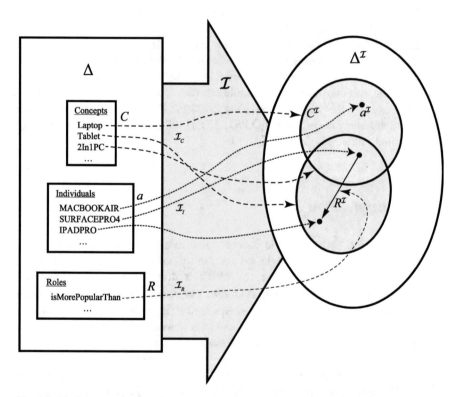

**Fig. 4.1**  The interpretation functions map concepts to subsets of the object domain, roles to binary
relations, and individuals to elements of the object domain

---

[6]The letter $\mathcal{I}$ of interpretation should not be confused with the $\mathcal{I}$ in description logic names, which
denotes inverse roles (although both the letter and the font are the same).

[7]The formal definition of the interpretation function varies from description logic to description
logic; the specific interpretation function for some of the most common and important description
logics is defined later in this chapter.

Datatypes are mapped into subsets of the datatype domain and data values are mapped into elements of the datatype domain.

The model-theoretic semantics of description logic concepts, individuals, roles, datatype constructors, and axioms based on these interpretations are detailed in the following sections.

### 4.1.4   DL Constructor Syntax and Semantics

Description logics utilize constructors such as atomic concepts and concept expressions, atomic roles and role expressions, complex role inclusions, and individual assertions, whose availability varies from description logic to description logic. Only the core formalisms are available in all description logics (e.g., concept inclusion, individual assertion). The formal meaning of description logic axioms is defined by their model-theoretic semantics (see Table 4.3).

$C \sqcap D$ is a concept comprising of all individuals that are individuals of both $C$ and $D$. $C \sqcup D$ contains individuals that are present in $C$ or $D$, or in both.

The existential quantification $\exists R.C$ is a concept that denotes the existence of domain individual $b$ such that domain individual $a$ is connected to $b$ via relation $R$, where $b$ is the element of the interpretation of $C$.

The universal quantification $\forall R.C$ denotes the set of domain individuals in which if domain individual $a$ is connected to some domain individual $b$ via relation $R$, then $b$ belongs to the interpretation of concept $C$.

Qualified at-least restriction refers to the domain elements for which there are at least $n$ individuals to which they are $R$-related and that are elements of the interpretation of $C$. Qualified at-most restriction refers to the domain elements for which a maximum of $n$ individuals exist to which they are $R$-related and that are elements of the interpretation of $C$.

Local reflexivity represents domain individuals that are $R$-related to themselves.

Role-value maps are very expressive concept constructors that relate the sets of role fillers[8] of the composition of role names (see Definition 4.1).

**Definition 4.1 (Role-Value Map)** If $R$, $S$ are role chains of the form $R_1 \circ \ldots \circ R_n$, where $R_n$ represents role names, then $R \subseteq S$ and $R = S$ are concepts and called *containment role-value maps* and *equality role-value maps*, respectively.

For example, the concept Person $\sqcap$ (friendOf $\circ$ hasHobby $=$ hasHobby) collects those persons who share hobbies with their friends.

Some constructors defined above for concepts are also available for roles, such as intersection, union, and complement; however, there are role-specific constructors as well. The most common role constructors are summarized in Table 4.4.

Table 4.5 summarizes the DL constructors of datatypes, data values, and their restrictions, which can complement the set of constructors of any description logic.

---

[8]Role fillers distinguish the function of concepts in a relationship. For example, individuals that belong to the Vehicle concept in the expression $\forall$hasWheels.Vehicle are fillers.

**Table 4.3** Syntax and semantics of concept and individual constructors

| Constructor | Syntax | Semantics |
|---|---|---|
| Atomic concept/ concept name | $A$ | $A^{\mathcal{I}}$ |
| Intersection/ conjunction | $C \sqcap D$ | $(C \sqcap D)^{\mathcal{I}} = C^{\mathcal{I}} \cap D^{\mathcal{I}}$ |
| Union/ disjunction | $C \sqcup D$ | $(C \sqcup D)^{\mathcal{I}} = C^{\mathcal{I}} \cup D^{\mathcal{I}}$ |
| Complement/ negation | $\neg C$ | $(\neg C)^{\mathcal{I}} = \Delta^{\mathcal{I}} \backslash C^{\mathcal{I}}$ |
| Existential quantification | $\exists R.C$ | $(\exists R.C)^{\mathcal{I}} = \{a \in \Delta^{\mathcal{I}} \mid \exists b.\langle a, b \rangle \in R^{\mathcal{I}} \wedge b \in C^{\mathcal{I}}\}$ |
| Universal quantification | $\forall R.C$ | $(\forall R.C)^{\mathcal{I}} = \{a \in \Delta^{\mathcal{I}} \mid \forall b.\langle a, b \rangle \in R^{\mathcal{I}} \rightarrow b \in C^{\mathcal{I}}\}$ |
| Top concept/ universal set/ tautology | $\top$ | $\Delta^{\mathcal{I}}$ |
| Bottom concept/ empty set/ contradiction | $\bot$ | $\emptyset$ |
| Individual | $a$ | $a^{\mathcal{I}} \in \Delta^{\mathcal{I}}$ |
| Nominal | $\{a, b, \ldots\}$ | $\{a, b, \ldots\}^{\mathcal{I}} = \{a^{\mathcal{I}}, b^{\mathcal{I}}, \ldots\}$ |
| Unqualified at-least restriction | $\geqslant nR$ | $(\geqslant nR)^{\mathcal{I}} = \{a \in \Delta^{\mathcal{I}} \mid \mid \{b \in \Delta^{\mathcal{I}} \mid \langle a, b \rangle \in R^{\mathcal{I}}\} \mid \geqslant n\}$ |
| Unqualified at-most restriction | $\leqslant nR$ | $(\leqslant nR)^{\mathcal{I}} = \{a \in \Delta^{\mathcal{I}} \mid \mid \{b \in \Delta^{\mathcal{I}} \mid \langle a, b \rangle \in R^{\mathcal{I}}\} \mid \leqslant n\}$ |
| Unqualified equivalence | $= nR$ | $(= nR)^{\mathcal{I}} = \{a \in \Delta^{\mathcal{I}} \mid \mid \{b \in \Delta^{\mathcal{I}} \mid \langle a, b \rangle \in R^{\mathcal{I}}\} \mid = n\}$ |
| Qualified at-least restriction | $\geqslant nR.C$ | $(\geqslant nR.C)^{\mathcal{I}} = \{a \in \Delta^{\mathcal{I}} \mid \mid \{b \in \Delta^{\mathcal{I}} \mid \langle a, b \rangle \in R^{\mathcal{I}} \wedge b \in C^{\mathcal{I}}\} \mid \geqslant n\}$ |
| Qualified at-most restriction | $\leqslant nR.C$ | $(\leqslant nR.C)^{\mathcal{I}} = \{a \in \Delta^{\mathcal{I}} \mid \mid \{b \in \Delta^{\mathcal{I}} \mid \langle a, b \rangle \in R^{\mathcal{I}} \wedge b \in C^{\mathcal{I}}\} \mid \leqslant n\}$ |
| Qualified equivalence | $= nR.C$ | $(= nR.C)^{\mathcal{I}} = \{a \in \Delta^{\mathcal{I}} \mid \mid \{b \in \Delta^{\mathcal{I}} \mid \langle a, b \rangle \in R^{\mathcal{I}} \wedge b \in C^{\mathcal{I}}\} \mid = n\}$ |
| Agreement | $u_1 \doteq u_2$ | $(u_1 \doteq u_2)^{\mathcal{I}} = \{a \in \Delta^{\mathcal{I}} \mid \exists b \in \Delta^{\mathcal{I}}.u_1^{\mathcal{I}}(a) = b = u_2^{\mathcal{I}}(a)\}$ |
| Disagreement | $u_1 \not\doteq u_2$ | $(u_1 \not\doteq u_2)^{\mathcal{I}} = \{a \in \Delta^{\mathcal{I}} \mid \exists b_1, b_2 \in \Delta^{\mathcal{I}}.u_1^{\mathcal{I}}(a) = b_1 \neq b_2 = u_2^{\mathcal{I}}(a)\}$ |
| Local reflexivity | $\exists R.Self$ | $(\exists R.Self)^{\mathcal{I}} = \{a \mid \langle a, a \rangle \in R^{\mathcal{I}}\}$ |
| Role-value map | $R \subseteq S$ $R = S$ | $(R \subseteq S)^{\mathcal{I}} = \{a \in \Delta^{\mathcal{I}} \mid \{\forall b.\langle a, b \rangle \in R^{\mathcal{I}} \rightarrow \langle a, b \rangle \in S^{\mathcal{I}}\}$ $(R = S)^{\mathcal{I}} = \{a \in \Delta^{\mathcal{I}} \mid \{\forall b.\langle a, b \rangle \in R^{\mathcal{I}} \leftrightarrow \langle a, b \rangle \in S^{\mathcal{I}}\}$ |

$C, D \in \mathbf{C}$ concepts, $A \in N_C$ concept name, $R, S \in \mathbf{R}$ roles, $a, b \in N_I$ individual names, $u_1, u_2$ chains of functional roles ($u_1 = f_1 \ldots f_m$, $u_2 = g_1 \ldots g_n$, where $n, m \geqslant 0$ and $f_i, g_j$ are features), $n$ nonnegative integer

**Table 4.4** Syntax and semantics of common role constructors

| Constructor | Syntax | Semantics |
|---|---|---|
| Atomic role/abstract role/role name | $R$ | $R^{\mathcal{I}} \subseteq \Delta^{\mathcal{I}} \times \Delta^{\mathcal{I}}$ |
| Universal role | $U$ | $U^{\mathcal{I}} \subseteq \Delta^{\mathcal{I}} \times \Delta^{\mathcal{I}}$ |
| Intersection/conjunction | $R \sqcap S$ | $(R \sqcap S)^{\mathcal{I}} = R^{\mathcal{I}} \cap S^{\mathcal{I}}$ |
| Union/disjunction | $R \sqcup S$ | $(R \sqcup S)^{\mathcal{I}} = R^{\mathcal{I}} \cup S^{\mathcal{I}}$ |
| Complement/negation | $\neg R$ | $(\neg R)^{\mathcal{I}} = \Delta^{\mathcal{I}} \times \Delta^{\mathcal{I}} \setminus R^{\mathcal{I}}$ |
| Inverse role | $R^-$ | $(R^-)^{\mathcal{I}} = \{\langle b, a\rangle \in \Delta^{\mathcal{I}} \times \Delta^{\mathcal{I}} \mid \langle a, b\rangle \in R^{\mathcal{I}}\}$ |
| Composition | $R \circ S$ | $R \circ S = R^{\mathcal{I}} \circ S^{\mathcal{I}}$ |
| Transitive closure | $R^+$ | $(R^+)^{\mathcal{I}} = \bigcup_{n \geq 1}(R^{\mathcal{I}})^n$ |
| Reflexive-transitive closure | $R^*$ | $(R^*)^{\mathcal{I}} = \bigcup_{n \geq 0}(R^{\mathcal{I}})^n$ |
| Role restriction | $R|_C$ | $(R|_C)^{\mathcal{I}} = R^{\mathcal{I}} \cap (\Delta^{\mathcal{I}} \times C^{\mathcal{I}})$ |
| Identity | $id(C)$ | $id(C) = \{\langle d, d\rangle \mid d \in C^{\mathcal{I}}\}$ |

**Table 4.5** Syntax and semantics of datatype and data value constructors

| Constructor | Syntax | Semantics |
|---|---|---|
| Datatype | $D$ | $D^{\mathcal{D}} \subseteq \Delta^{\mathcal{I}}_{\mathcal{D}}$ |
| Datatype role | $U$ | $U^{\mathcal{I}} \subseteq \Delta^{\mathcal{I}} \times \Delta^{\mathcal{I}}_{\mathcal{D}}$ |
| Data value | $v$ | $v^{\mathcal{I}} = v^{\mathcal{D}}$ |
| Data enumeration | $\{v_1, \ldots\}$ | $\{v_1, \ldots\}^{\mathcal{I}} = \{v^{\mathcal{I}}, \ldots\}$ |
| Existential datatype restriction | $\exists U.D$ | $(\exists U.D)^{\mathcal{I}} = \{x \mid \exists y.\langle x, y\rangle \in U^{\mathcal{I}} \text{ and } y \in D^{\mathcal{D}}\}$ |
| Universal datatype value restriction | $\forall U.D$ | $(\forall U.D)^{\mathcal{I}} = \{x \mid \forall y.\langle x, y\rangle \in U^{\mathcal{I}} \to y \in D^{\mathcal{D}}\}$ |
| At-least datatype restriction | $\geq nU$ | $(\geq nU)^{\mathcal{I}} = \{x \mid \sharp(\{y.\langle x, y\rangle \in U^{\mathcal{I}}\}) \geq n\}$ |
| At-most datatype restriction | $\leq nU$ | $(\leq nU)^{\mathcal{I}} = \{x \mid \sharp(\{y.\langle x, y\rangle \in U^{\mathcal{I}}\}) \leq n\}$ |

$v$, data value; $x,y$, individuals; $\sharp$, set cardinality

The datatype and data value constructors are not available in all description logics. They are denoted by the $^{(\mathcal{D})}$ superscript at the end of the name of all description logics that support them, as seen in $\mathcal{SHOIN}^{(\mathcal{D})}$ and $\mathcal{SROIQ}^{(\mathcal{D})}$, for example.

### 4.1.5 DL Axiom Syntax and Semantics

Formal description logic statements employ *axioms* to express concept inclusion, role membership, individual assertion, complex role inclusion, and so on. The syntax and semantics of description logic axioms is summarized in Table 4.6.

The satisfaction of axioms is determined by the lifted interpretation function $\mathcal{I}$. For example, a *general concept inclusion* of the form $C \sqsubseteq D$ (Definition 4.2) is satisfied by $\mathcal{I}$ if every instance of $C$ is also an instance of $D$.

**Table 4.6** Syntax and semantics of common description logic axioms

| Axiom | Syntax | Semantics |
|---|---|---|
| Concept inclusion | $C \sqsubseteq D$ | $C^{\mathcal{I}} \subseteq D^{\mathcal{I}}$ |
| Concept equivalence | $C \equiv D$ | $C^{\mathcal{I}} = D^{\mathcal{I}}$ |
| Individual assertion | $C(a)$ or $a : C$ | $a^{\mathcal{I}} \in C^{\mathcal{I}}$ |
| Role membership | $R(a, b)$ or $(a, b) : R$ | $\langle a^{\mathcal{I}}, b^{\mathcal{I}} \rangle \in R^{\mathcal{I}}$ |
| Individual equality | $a \approx b$ | $a^{\mathcal{I}} = b^{\mathcal{I}}$ |
| Individual inequality | $a \not\approx b$ | $a^{\mathcal{I}} \neq b^{\mathcal{I}}$ |
| Object role inclusion | $R \sqsubseteq S$ | $R^{\mathcal{I}} \subseteq S^{\mathcal{I}}$ |
| Datatype role inclusion | $U \sqsubseteq V$ | $U^{\mathcal{I}} \sqsubseteq V^{\mathcal{I}}$ |
| Role equivalence | $R \equiv S$ | $R^{\mathcal{I}} = S^{\mathcal{I}}$ |
| Complex role inclusion | $R_1 \circ R_2 \sqsubseteq S$ | $R_1^{\mathcal{I}} \circ R_2^{\mathcal{I}} \subseteq S^{\mathcal{I}}$ |
| Role assertion | $Disjoint(R, S)$ $Transitive(R)$ ... | $R^{\mathcal{I}} \cap S^{\mathcal{I}} = \emptyset$ $R^{\mathcal{I}} = (R^{\mathcal{I}})^{+}$ ... |

**Definition 4.2 (General Concept Inclusion, GCI)** A general concept inclusion is an axiom of the form $\langle C \sqsubseteq D \rangle$, where $C, D$ are concepts. If both $\langle C \sqsubseteq D \rangle$ and $\langle D \sqsubseteq C \rangle$ hold, the axiom is written as concept equivalence of the form $C \equiv D$. If $C$ is an atomic concept in a GCI, the axiom is called a *primitive GCI*.

A *complex role inclusion axiom* of the form $R_1 \circ \dots \circ R_n \sqsubseteq S$ (see Definition 4.3) holds in $\mathcal{I}$ if for every sequence $\delta_0, \dots, \delta_n \in \Delta^{\mathcal{I}}$ for which $\langle \delta_0, \delta_1 \rangle \in R_1^{\mathcal{I}}, \dots,$ $\langle \delta_{n-1}, \delta_n \rangle \in R_n^{\mathcal{I}}$ holds, $\langle \delta_0, \delta_n \rangle \in R_n^{\mathcal{I}}$ is also satisfied.

**Definition 4.3 (Role Inclusion Axiom, RIA)** A *generalized role inclusion* is an axiom of the form $R \sqsubseteq S$, where $R$ and $S$ are roles. Role inclusion axioms that include role composition of the form $R_1 \circ \dots \circ R_n \sqsubseteq S$ are called complex role inclusion axioms.

The role inclusion axioms of a knowledge base form a role hierarchy (see Definition 4.4).

**Definition 4.4 (Role Hierarchy)** A role hierarchy $R_h$ is a finite set of role inclusion axioms.

To prevent cyclic dependencies in role hierarchies, regularity can be set as a restriction (see Definition 4.5).

**Definition 4.5 (Regular Role Hierarchy)** A role hierarchy $R_h$ is regular if there is a regular order $\prec$ on roles such that each role inclusion axiom in $R_h$ is $\prec$-regular.

Based on a role hierarchy, models can be defined (see Definition 4.6).

**Definition 4.6 (Model)** An interpretation $\mathcal{I}$ is a model of a role hierarchy $R_h$ if it satisfies all role inclusion axioms in $R_h$, i.e., $\mathcal{I} \models R_h$.

Restrictions on role interpretations enforce the interpretation of roles to satisfy certain properties, such as functionality (see Definition 4.7), transitivity (see Definition 4.8), and disjointness (see Definition 4.9).

**Definition 4.7 (Functional Role)** Consider a subset $N_F$ of the set of role names $N_R$, whose elements are called features. An interpretation must map each feature $f$ to functional binary relations $f^{\mathcal{I}} \subseteq \Delta^{\mathcal{I}} \times \Delta^{\mathcal{I}}$, i.e., relations satisfying $\forall a,b,c.f^{\mathcal{I}}(a, b) \wedge f^{\mathcal{I}}(a,c) \rightarrow b = c$. When functional relations are considered partial functions, $f^{\mathcal{I}}(a,b)$ is written as $f^{\mathcal{I}}(a) = b$.

**Definition 4.8 (Transitive Role)** Consider a subset $N_{R+}$ of $N_R$ called the set of transitive roles. An interpretation must map transitive roles $R \in N_{R+}$ to transitive binary relations $R^{\mathcal{I}} \subseteq \Delta^{\mathcal{I}} \times \Delta^{\mathcal{I}}$.

**Definition 4.9 (Role Disjointness)** A role disjointness statement $Disjoint(R, S)$ is true in $\mathcal{I}$ if every two domain individuals $a, b \in \Delta^{\mathcal{I}}$ connected via an $R$-relation are not connected via an $S$-relation.

Such role characteristics are written in the form of role assertions (see Definition 4.10).

**Definition 4.10 (Role Assertion)** For roles $R,S \neq U$, the assertions *Reflexive* (R), *Irreflexive* (R), *Symmetric* (R), *Transitive* (R), and *Disjoint* (R,S) are called role assertions,[9] where for each interpretation $\mathcal{I}$ and all $x, y, z \in \Delta^{\mathcal{I}}$

$$\mathcal{I} \vDash Symmetric(R) \text{ if } \langle x, y \rangle \in R^{\mathcal{I}} \text{ implies } \langle y, x \rangle \in R^{\mathcal{I}}$$

$$\mathcal{I} \vDash Transitive(R) \text{ if } \langle x, y \rangle \in R^{\mathcal{I}} \text{ and } \langle y, z \rangle \in R^{\mathcal{I}} \text{ imply } \langle x, z \rangle \in R^{\mathcal{I}}$$

$$\mathcal{I} \vDash Reflexive(R) \text{ if } Diag^{\mathcal{I}} \in R^{\mathcal{I}}$$

$$\mathcal{I} \vDash Irreflexive(R) \text{ if } R^{\mathcal{I}} \cap Diag^{\mathcal{I}} = \emptyset$$

$$\mathcal{I} \vDash Disjoint(R, S) \text{ if } R^{\mathcal{I}} \cap S^{\mathcal{I}} = \emptyset$$

where $Diag^{\mathcal{I}}$ is defined as the set $\{ \langle x, x \rangle \mid x \in \Delta^{\mathcal{I}} \}$ and $R^{\mathcal{I}} \downarrow := \{ \langle x, x \rangle \mid \exists y \in \Delta^{\mathcal{I}}.\langle x, y \rangle \in R^{\mathcal{I}} \}$.

The source and target individuals for a role can be defined as the *role domain* and the *role range*, respectively (see Definitions 4.11 and 4.12).

**Definition 4.11 (Role Domain)** A role $R$ has domain $C$ in an interpretation $\mathcal{I}$ if all the source individuals of the relation associated with $R$ are instances of concept $C$.

**Definition 4.12 (Role Range)** An object role $R$ has range $C$ in an interpretation $\mathcal{I}$ if all target individuals of the relation associated with $R$ are instances of concept $C$. A datatype role $R$ has range $D$ in an interpretation $\mathcal{I}$ if all target individuals of the

---

[9]In the DL literature, role assertions are often abbreviated by the first three letters, i.e., $Sym(R)$, $Tra(R)$, $Ref(R)$, $Irr(R)$, and $Dis(R)$. They are written in full throughout this book for clarity.

relation associated with $R$ are instances of data range $D$, which is either an RDF datatype, a typed or untyped RDF literal, or an enumerated datatype defining a range of permissible data values.

Domain restrictions are constraint axioms of the form $domain(R,C)$, and restrict the domain of role $R$ to be concept $C$. Range restrictions are written as axioms of the form $range(R,C)$, and restrict the range of role $R$ to be concept $C$.

The third type of constraint axioms is the disjoint restriction, which restricts the concept names $C$ and $D$ to be disjoint.

## 4.1.6  TBox, ABox, and RBox

The different types of description logic statements can be grouped as follows: terminological knowledge (*TBox*), assertional knowledge (*ABox*), and relational statements (*RBox*). TBox statements describe a conceptualization, a set of concepts, and properties for these concepts; the ABox collects facts, i.e., statements about individuals belonging to the concepts formalized in the TBox; and the RBox collects the role hierarchy and complex role inclusion axioms.

Terminological knowledge can be added to ontologies by defining the relationship of classes and properties as subclass axioms and subproperty axioms, respectively, and concept equivalence axioms in the TBox (see Definition 4.13).

**Definition 4.13 (TBox)** A TBox $\mathcal{T}$ is a finite collection of general concept inclusion axioms of the form $C \sqsubseteq D$, concept equivalence axioms of the form $C \equiv D$, and constraints, where $C$ and $D$ are concepts.[10]

For example, TBox axioms can express that live action is a movie type or narrators are equivalent to lectors, as shown in Table 4.7.

Individuals and their relationships are represented by ABox axioms. An individual assertion in an ABox can be a concept assertion, a role assertion, a negated role assertion, an equality statement, or an inequality statement (see Definition 4.14).

**Definition 4.14 (ABox)** An ABox $\mathcal{A}$ is a finite collection of concept membership axioms of the form $C(a)$, role membership axioms of the form $R(a, b)$, negated role membership axioms of the form $\neg R(a, b)$, individual equality axioms of the form $a \approx b$, and individual inequality axioms of the form $a \not\approx b$, where $a, b \in N_I$ are individual names, $C \in \mathbf{C}$ denotes concept expressions, and $R \in \mathbf{R}$ denotes roles, all of which are demonstrated in Table 4.8.[11]

---

[10]The $\mathcal{T}$ that stands for TBox is one of the exceptions that do not correspond to mathematical constructors from Table 4.2, but is considered the de facto standard annotation for TBoxes in the DL literature.

[11]While written with the CMSY10 font, the $\mathcal{A}$ that stands for ABox has nothing to do with the mathematical constructors of $\mathcal{AL}$ in Table 4.2, but is considered the de facto standard annotation for ABoxes in the DL literature.

**Table 4.7**  Expressing terminological knowledge with TBox axioms

| DL syntax | Turtle syntax |
|---|---|
| liveAction ⊑ Movie | `:liveAction rdfs:subClassOf :Movie .` |
| Narrator ≡ Lector | `:Narrator owl:equivalentClass :Lector .` |
| Movie ≡ LiveAction ⊔ Animation ⊔ Cartoon | `:Movie owl:unionOf (:LiveAction :Animation :Cartoon)` |
| CartoonWithMotionPicture ≡ Cartoon ⊓ MotionPicture | `:CartoonWithMotionPicture owl:intersectionOf (:Cartoon :MotionPicture)` |
| ¬Cartoon | `:LiveAction owl:complementOf :Cartoon .` |
| Cast ≡ ∀hasMember.Actor | `:Cast owl:equivalentClass [ a owl:Restriction ; owl:onProperty :hasMember ; owl:allValuesFrom :Actor ].` |
| Director ≡ ∃directorOf.Movie | `:Director owl:equivalentClass [ a owl:Restriction ; owl:onProperty :directorOf ; owl:someValuesFrom :Movie ] .` |

**Table 4.8**  Asserting individuals with ABox axioms

| DL syntax | Turtle syntax |
|---|---|
| computerAnimation(ZAMBEZIA) | `:Zambezia a :computerAnimation .` |
| directedBy(UNFORGIVEN, CLINTEASTWOOD) | `:Unforgiven :directedBy :ClintEastwood .` |
| 房仕龍 ≈ JACKIECHAN | `:房仕龍 owl:sameAs :JackieChan .` |
| ROBINWILLIAMS ≉ ROBBIEWILLIAMS | `:RobinWilliams owl:differentFrom :RobbieWilliams .` |

The distinction between the TBox and the ABox is not always significant; however, it is useful when describing and formulating decision procedures for various description logics. Reasoners might process the TBox and the ABox separately, because some key inference problems are affected by either the TBox or the ABox only, but not both. For example, classification can be performed using the TBox alone, while instance checking requires the ABox only.[12] Also, the complexity of the TBox helps determine the performance of decision procedures for a particular description logic, independently from the ABox. Reasoning in the presence of a TBox is significantly harder than reasoning without a TBox, especially if the TBox contains terminological cycles (*cyclic TBox*),[13] i.e., a recursive concept inclusion, e.g., Movie $\sqsubseteq$ $\forall$directedBy.Director, or one or more mutually recursive concept inclusions, such as { Movie $\sqsubseteq$ $\exists$directedBy.Director, Director $\sqsubseteq$ $\exists$directorOf.Movie }. Moreover, concept definitions from a TBox might be reused across ontologies that do not share the same ABox.

Most multimedia ontologies define terminological and assertional axioms only. Beyond ABox and TBox axioms, however, *role box axioms* can also be defined to collect all statements related to roles, and describe role characteristics and the interdependencies between roles, as defined in Definition 4.15 [109].

**Definition 4.15 (RBox)**   A role box $\mathcal{R}$ (RBox)[14] consists of (1) a role hierarchy, i.e., a finite collection of generalized role inclusion axioms of the form $R \sqsubseteq S$, role equivalence axioms of the form $R \equiv S$, complex role inclusions of the form $R_1 \circ R_2 \sqsubseteq S$, and (2) a set $R_a$ of role assertions, such as reflexive role axioms of the form *Reflexive* $(R)$, irreflexive role axioms of the form *Irreflexive* $(R)$, symmetric role axioms of the form *Symmetric* $(R)$, transitive role axioms of the form *Transitive* $(R)$, and role disjointness axioms of the form *Disjoint* $(R, S)$, where $R$ and $S$ are roles.[15]

Some examples for role box axioms are shown in Table 4.9.

**Table 4.9**   Modeling relationships between roles with RBox axioms

| DL syntax | Turtle syntax |
|---|---|
| hasChild $\circ$ hasChild $\sqsubseteq$ hasGrandchild | `:hasGrandchild owl:propertyChainAxiom` `(:hasChild :hasChild) .` |
| Disjoint(parentOf, childOf) | `[] a owl:AllDisjointProperties ;` `owl:members (:parentOf :childOf) .` |
| basedOn $\circ$ basedOn $\sqsubseteq$ basedOn | `vidont:basedOn a` `owl:TransitiveProperty .` |

---

[12]For description logics that support full negation of concepts, the instance checking can be reduced to the problem of deciding whether ABox $\mathcal{A} \cup \{\neg C(a)\}$ is inconsistent (with reference to TBox $\mathcal{T}$).

[13]A TBox that does not contain terminological cycles is called an *acyclic TBox*.

[14]While written with the CMSY10 font, the $\mathcal{R}$ that stands for RBox should not be confused with the mathematical constructors of $\mathcal{R}$ in Table 4.2, but is considered the de facto standard annotation for RBoxes in the DL literature.

[15]This is the definition of the $\mathcal{SROIQ}$ RBox, which is a superset of the RBox of less expressive description logics, whose definitions are inherently redundant and hence they are omitted here.

The RBox is available in very expressive description logics only, such as $\mathcal{SHIQ}$, $\mathcal{SHOIN}$, and $\mathcal{SROIQ}$, but all DL-based ontologies have a TBox and most of them an ABox.

A TBox, an ABox, and an RBox together form a description logic knowledge base (see Definition 4.16), which corresponds to OWL 2 ontologies.

**Definition 4.16 (Knowledge Base)**   A DL knowledge base $\mathcal{K}$ is a triple $\langle \mathcal{T}, \mathcal{A}, \mathcal{R} \rangle$,[16] where $\mathcal{T}$ is a set of terminological axioms (TBox), $\mathcal{A}$ is a set of assertional axioms (ABox), and $\mathcal{R}$ is a role box (RBox).[17]

## *4.1.7   Relation to Other Logics*

Many description logics are decidable fragments of first-order logic[18] and are fully fledged logics with formal semantics. Description logics are closely related to propositional modal logic, which extends classical propositional and predicate logic to include operators expressing modality in the form of necessity, possibility, impossibility, and related notions, as well as *guarded fragments*, which generalize modal quantification through finding relative patterns of quantification.

### 4.1.7.1   Relation to First-Order Logic

First-order logic, also known as first-order predicate calculus (FOPC), is a logic that supports variable symbols, such as $x$, $y$, connectives, such as not (~), and (^), or (V), implies (=>), if and only if (<=>), and universal and existential quantifiers (A and E, respectively). User-defined primitives include constant symbols, i.e., the individuals of the represented world (e.g., 5), function symbols, i.e., mapping between individuals (e.g., *colorOf(Sky)* = *blue*), and predicate symbols, which map individuals to truth values, such as *color(Grass, green)*.

FOL sentences consist of terms and atoms. A term, which denotes a real-world individual, is a constant symbol, a variable symbol, or an $n$-place function of $n$ terms. For example, $x$ and $f(x_1, \ldots, x_n)$ are terms, where each $x_i$ is a term. An atom is either an $n$-place predicate of $n$ terms or connectives (~$P$, $P \vee Q$, $P \wedge Q$, $P => Q$, $P <=>$ $Q$, where $P$ and $Q$ are atoms). FOL atoms have Boolean values. A sentence is an atom or a combination of a quantifier and a variable (e.g., $(Ax)P$, $(Ex)P$, where $P$ is a

---

[16]The letter $\mathcal{K}$ is one of the very few letters that are written in the CMSY10 font in the DL literature, yet does not correspond to mathematical constructors from Table 4.2.

[17]In the literature, the knowledge base definition often includes the TBox-ABox pair only, which corresponds to OWL ontologies, and does not include an RBox. For very expressive description logics, the definition includes the RBox as well.

[18]There are some exceptions, such as $\mathcal{CIQ}$, for example, which is not a fragment of FOL due to its transitive reflexive closure constructor.

sentence and $x$ is a variable). A well-formed formula is a sentence without free variables, where all variables are bound by universal or existential quantifiers.

Based on the relations between FOL and DL, a transparent translation is feasible between FOL and most DL formalisms, as demonstrated in Table 4.10.

**Table 4.10** Transparent translation between FOL and DL

| First-order logic | Description logic |
| --- | --- |
| $A(x)$ | $A$ |
| $C(a)$ | $C(a)$ |
| $A \approx B$ | $A \equiv B$ |
| $\neg C(x)$ | $\neg C$ |
| $C(x) \wedge D(x)$ | $C \sqcap D$ |
| $C(x) \vee D(x)$ | $C \sqcup D$ |
| $\forall x(C(x) \rightarrow D(x))$ | $C \sqsubseteq D$ |
| $R(a,b)$ | $R(a,b)$ |
| $\forall x \forall y(R(x,y) \rightarrow S(x,y))$ | $R \sqsubseteq S$ |
| $\exists y(R(x,y) \wedge C(y))$ | $\exists R.C$ |
| $\forall x \forall y\, (R(x,y) \rightarrow R(y,z) \rightarrow R(x,z))$ | $R \circ R \sqsubseteq R$ |

Despite the feasibility of direct translation between FOL and DL, guaranteeing complete and terminating reasoning requires a different transformation, such as the *structural transformation* [110]. The structural transformation is based on a conjunction normal form, which replaces FOL subformulae with new predicates, for which it also provides definitions. A major advantage of the structural transformation is that it avoids the exponential size growth of the clauses.

While many description logics are decidable fragments of first-order logic, two-variable logic, or guarded logic [111], some description logics have more features than first-order logic. Similar to description logics, first-order logic can also be used for the formal description of multimedia contents [112].

### 4.1.7.2  Relation to Modal Logic

Description logics are related to, but developed independently from, *modal logic* (ML). Many description logics are syntactic variants of modal logic, so many logical representations expressed in modal logic correspond to description logic formalisms [113], as demonstrated in Table 4.11.

In modal logic, an object usually corresponds to a possible world, a concept to a modal proposition, and a role-bounded quantifier to a modal operator. Operations on roles (e.g., composition) correspond to the modal operations used in dynamic logic.

**Table 4.11**  Examples for modal logic to description logic translation

| Modal logic | Description logics |
|---|---|
| $M, x \vDash \neg\phi$ iff $M, x \nvDash \phi$ <br> $(\neg\phi)^M = \{ x \mid x \notin \phi^M \}$ | $(\neg C)^{\mathcal{I}} = \Delta^{\mathcal{I}} \setminus C^{\mathcal{I}}$ |
| $M, x \vDash \phi_1 \wedge \phi_2$ iff $M, x \vDash \phi_1$ and $M, x \vDash \phi_2$ <br> $(\phi_1 \wedge \phi_2)^M = \phi_1^M \cap \phi_2^M$ | $(C \sqcap D)^{\mathcal{I}} = C^{\mathcal{I}} \cap D^{\mathcal{I}}$ |
| $M, x \vDash \phi_1 \vee \phi_2$ iff $M, x \vDash \phi_1$ and $M, x \vDash \phi_2$ <br> $(\phi_1 \vee \phi_2)^M = \phi_1^M \cup \phi_2^M$ | $(C \sqcup D)^{\mathcal{I}} = C^{\mathcal{I}} \cup D^{\mathcal{I}}$ |
| $M, x \vDash \diamond\phi$ iff $\exists y((x,y) \in R$ and $M, y \vDash \phi)$ <br> $(\diamond\phi)^M = \{ x \mid \exists y((x,y) \in R$ and $y \in \phi^M)\}$ | $(\exists R.C)^{\mathcal{I}} = \{x \in \Delta^{\mathcal{I}} \mid \exists y.(x, y) \in R^{\mathcal{I}} \wedge y \in C^{\mathcal{I}}\}$ |
| $M, x \vDash \Box\phi$ iff $\forall y((x,y) \in R \rightarrow M, y \vDash \phi)$ <br> $(\Box\phi)^M = \{ x \mid \forall y((x,y) \in R \rightarrow y \in \phi^M)\}$ | $(\forall R.C)^{\mathcal{I}} = \{x \in \Delta^{\mathcal{I}} \mid \forall y.(x, y) \in R^{\mathcal{I}} \rightarrow y \in C^{\mathcal{I}}\}$ |

## 4.2  Description Logic Families

The balance between the two main aspects of description logic implementations, reasoning complexity and expressivity, depends on the application. For lightweight description logics, such as $\mathcal{ALC}$ or $\mathcal{EL}^{++}$, the most common reasoning problems can be implemented in (sub)polynomial time. For other applications, in which expressivity has a priority over reasoning complexity, very expressive description logics are used, such as $\mathcal{SHOIN}$ or $\mathcal{SROIQ}$.

### 4.2.1  *ALC and the Basic Description Logics*

The *Attributive Language* ($\mathcal{AL}$) family of description logics allows atomic negation, concept intersection, universal restrictions, and limited existential quantification.

A core $\mathcal{AL}$-based description logic is the *Attributive (Concept) Language with Complements* ($\mathcal{ALC}$), in which, unlike $\mathcal{AL}$, the complement of any concept is allowed, not only the complement of atomic concepts [114]. From the permissible constructors point of view, $\mathcal{ALC}$ would be equivalent to $\mathcal{ALUE}$, although the latter name is not used.

$\mathcal{ALC}$ concept expressions can include concept names, concept intersection, concept union, complement, existential and universal quantifiers, and individual names (see Definition 4.17).

**Definition 4.17 ($\mathcal{ALC}$ Concept Expression)**  Let $N_C$ be a set of concept names and $N_R$ a set of role names. The set of $\mathcal{ALC}$ concept expressions is the smallest set such that $\top$, $\bot$, and every concept name $A \in N_C$ is an $\mathcal{ALC}$ concept description, and if $C$ and $D$ are $\mathcal{ALC}$ concept descriptions and $R \in N_R$, then $C \sqcap D$, $C \sqcup D$, $\neg C$, $\forall R.C$, and $\exists R.C$ are also $\mathcal{ALC}$ concept descriptions.

$\mathcal{ALC}$ interpretations are defined in Definition 4.18.

**Definition 4.18 ($\mathcal{ALC}$ Interpretation)**  A terminological $\mathcal{ALC}$ interpretation $\mathcal{I} = (\Delta^{\mathcal{I}}, \cdot^{\mathcal{I}})$ over a signature $(N_C, N_R, N_I)$ consists of a nonempty set $\Delta^{\mathcal{I}}$ (domain) and an interpretation function $^{\mathcal{I}}$, which maps each individual $a$ to an element $a^{\mathcal{I}} \in \Delta^{\mathcal{I}}$, each concept to a subset of $\Delta^{\mathcal{I}}$, and each role name to a subset of $\Delta^{\mathcal{I}} \times \Delta^{\mathcal{I}}$, such that for all $\mathcal{ALC}$ concepts $C$ and $D$ and all role names $R$, $\top^{\mathcal{I}} = \Delta^{\mathcal{I}}$, $\bot^{\mathcal{I}} = \emptyset$, $(C \sqcap D)^{\mathcal{I}} = C^{\mathcal{I}} \cap D^{\mathcal{I}}$, $(C \sqcup D)^{\mathcal{I}} = C^{\mathcal{I}} \cup D^{\mathcal{I}}$, $\neg C = \Delta^{\mathcal{I}} \backslash C^{\mathcal{I}}$, $(\exists R.C)^{\mathcal{I}} = \{x \in \Delta^{\mathcal{I}} \mid \text{there is some } y \in \Delta^{\mathcal{I}} \text{ with } \langle x, y \rangle \in R^{\mathcal{I}} \text{ and } y \in C^{\mathcal{I}}\}$, and $(\forall R.C)^{\mathcal{I}} = \{x \in \Delta^{\mathcal{I}} \mid \text{for all } y \in \Delta^{\mathcal{I}}, \text{if } \langle x, y \rangle \in R^{\mathcal{I}}, \text{then } y \in C^{\mathcal{I}}\}$.

Since $\mathcal{ALC}$ is sufficiently expressive to support many fundamental constructors for web ontologies, it serves as the basis for many slightly more expressive description logics (e.g., $\mathcal{ALCHIF}$, $\mathcal{ALCIQ}$) and most very expressive description logics ($\mathcal{SHIF}$, $\mathcal{SHIQ}$, $\mathcal{SHOIN}$, $\mathcal{SROIQ}$, etc.). Several fragments of $\mathcal{ALC}$ are also used, such as logics in the $\mathcal{EL}$, $DL$-$Lite$, and $\mathcal{FL}$ families, which will be discussed in the following sections, that usually restrict the use of Boolean operators and quantifiers, and so their reasoning problems are often easier than that of $\mathcal{ALC}$.

### 4.2.2 The $\mathcal{EL}$ Family of Description Logics

The $\mathcal{EL}$ family of description logics allows concept intersection and existential restrictions of full existential quantification. $\mathcal{EL}$ is the formal basis of the EL profile of OWL 2. $\mathcal{EL}$ and its tractable[19] extensions are suitable for managing the large terminological knowledge bases commonly found in health sciences and bioinformatics. Such knowledge bases often require classification, which is done by computing all entailed (implicit) subsumption relations between atomic concepts [115] (see later in Chap. 6).

The names of $\mathcal{EL}$ description logics follow the general naming conventions of description logics, except that $\mathcal{ELRO}$ is sometimes called $\mathcal{EL}^{++}$ (which corresponds to OWL 2 EL). The benefit of $\mathcal{EL}^{++}$ is that it is expressive enough to reduce standard reasoning tasks, such as concept satisfiability, ABox consistency, and instance checking, to the subsumption problem [116], all of which will be detailed later in Chap. 6.

### 4.2.3 The DL-Lite Family of Description Logics

While tractable reasoning and efficient query answering are not key factors for small knowledge bases and ontologies, they are crucial when dealing with thousands of concepts and millions of individuals. Similar to the $\mathcal{EL}$ family, the $DL$-$Lite$ family of description logics was designed to provide core constructors for ontology engineering with low computational complexity. A common feature of the $\mathcal{EL}$ and the $DL$-$Lite$ family is that disjunction and universal restrictions are not supported. One of the implementations is OWL 2 QL, which is based on $DL$-$Lite$.

---

[19]Decidable in polynomial time.

Members of the *DL-Lite* family are based on the original *DL-Lite* description logic, which can be defined briefly as follows. *DL-Lite* concepts are defined as $B ::= A \mid \exists R \mid \exists R^-$, $C ::= B \mid \neg B \mid C_1 \sqcap C_2$, where $A$ denotes an atomic concept and $R$ denotes an atomic role, and $B$ denotes a basic concept that can be either an atomic concept, a construct with unified existential quantification on roles, or a concept that involves an inverse role. $C$ denotes a general concept. In *DL-Lite*, negation is allowed for basic concepts only. A *DL-Lite* TBox consists of inclusion assertions of the form $B \sqsubseteq C$ and functionality assertions of the form (funct $R$), (funct $R^-$). A *DL-Lite* ABox allows membership assertions of the form $B(a)$, $R(a,b)$, where $a$ and $b$ are constants, the object denoted by $a$ is an instance of the basic concept $B$, and the pair of objects denoted by $(a,b)$ is an instance of the role $R$.

One of the strengths of *DL-Lite* is that it allows very powerful querying compared to other, often much more expressive, description logics, in which only the concept membership or role membership can be queried. *DL-Lite* supports arbitrary complex conjunctive queries (see Definition 4.20).

**Definition 4.20 (Conjunctive Query)**  A conjunctive query (CQ) $q$ over a knowledge base $\mathcal{K}$ is an expression of the form $q(\vec{x}) \leftarrow \exists \vec{y}.\mathrm{conj}(\vec{x}, \vec{y})$, where $\vec{x}$ denotes distinguished variables, $\vec{y}$ denotes existentially quantified variables (nondistinguished variables), and $\mathrm{conj}(\vec{x}, \vec{y})$ is a conjunction of atoms of the form $B(z)$, or $R$ $(z_1, z_2)$, where $B$ and $R$ denote a basic concept and a role in $\mathcal{K}$, and $z$, $z_1$, $z_2$ are constants in $\mathcal{K}$ or variables in $\vec{x}$ or $\vec{y}$.

Interpretations in *DL-Lite* are defined as shown in Definition 4.21.

**Definition 4.21 (*DL-Lite* Interpretation)**  A *DL-Lite* interpretation $\mathcal{I} = (\Delta, ^{\mathcal{I}})$ consists of a first-order structure over $\Delta$ with an interpretation function $^{\mathcal{I}}$ such that $A^{\mathcal{I}} \subseteq \Delta, (\neg B)^{\mathcal{I}} = \Delta \backslash B^{\mathcal{I}}, (C_1 \sqcap C_2)^{\mathcal{I}} = C_1^{\mathcal{I}} \cap C_2^{\mathcal{I}}, R^{\mathcal{I}} \subseteq \Delta \times \Delta, (\exists R)^{\mathcal{I}} = \{ c \mid \exists c'.(c.c') \in R^{\mathcal{I}} \}$, and $(\exists R^-)^{\mathcal{I}} = \{ c \mid \exists c'.(c.c') \in R^{\mathcal{I}} \}$.

Different approaches to extend the original *DL-Lite* logic include the addition of Boolean connectives and number restrictions to concept constructs, allowing role hierarchies, role disjointness, symmetry, asymmetry, reflexivity, irreflexivity, and transitivity constraints, and adopting or dropping the unique name assumption. These approaches resulted in description logics such as $DL\text{-}Lite_{core}$, $DL\text{-}Lite_{F,\sqcap}$, $DL\text{-}Lite_{R,\sqcap}$, $DL\text{-}Lite_A$, $DL\text{-}Lite_{bool}$, $DL\text{-}Lite_{krom}$,[20] $DL\text{-}Lite_{horn}$, and their variants. For example, $DL\text{-}Lite_{horn}^F$ is a Horn extension of OWL 2 QL. $DL\text{-}Lite_{bool}^N$ contains object names ($a_0, a_1, \ldots$), concept names ($A_0, A_1, \ldots$), and role names ($P_0, P_1, \ldots$). $DL\text{-}Lite_{bool}^N$ roles are defined as $R ::= P_k \mid P_k^-$, basic concepts as $B ::= \bot \mid A_k \mid \geqslant qR$, where $q$ is a positive integer, and concepts as $C ::= B \mid \neg C \mid C_1 \sqcap C_2$. $DL\text{-}Lite_{bool}^N$ has two sublanguages, both of which restrict Boolean operators on concepts: $DL\text{-}Lite_{krom}^N$, whose TBoxes can only have concept inclusions of the form $B_1 \sqsubseteq$

---

[20]The Krom fragment of first-order logic consists of all formulae in *prenex normal form* (i.e., written as a string of quantifiers (prefix) followed by a quantifier-free part (matrix)), whose quantifier-free part is a conjunction of binary clauses.

$B_2, B_1 \sqsubseteq \neg B_2$, or $\neg B_1 \sqsubseteq B_2$, and $DL\text{-}Lite_{core}^{N}$, which can only have concept inclusions of the form $B_1 \sqsubseteq B_2$ or $B_1 \sqcap B_2 \sqsubseteq \top$.

## 4.2.4  Frame-Based Description Logics ($\mathcal{FL}$)

$\mathcal{FL}$ description logics allow concept intersection, universal restrictions, limited existential quantification, and role restrictions. $\mathcal{FL}^-$ is a sublanguage of $\mathcal{FL}$, in which role restriction is not allowed, i.e., the equivalent of $\mathcal{AL}$ without atomic negation. $\mathcal{FL}_0$ is a sublanguage of $\mathcal{FL}^-$, in which limited existential quantification is not allowed.

## 4.2.5  The $\mathcal{SH}$ Family of Description Logics

The common feature of description logics in the $\mathcal{SH}$ family is the support of the mathematical constructors of $\mathcal{ALC}$ (atomic negation, concept intersection, universal restrictions, limited existential quantification, subclass relationships, equivalence, conjunction, disjunction, negation, property restrictions, tautology, contradiction), complemented by transitive roles and role hierarchy. Some of the most common $\mathcal{SH}$-based description logics are $\mathcal{SHIF}, \mathcal{SHIQ}, \mathcal{SHOQ}, \mathcal{SHOIN}$, and $\mathcal{SROIQ}$, each of which adds a different set of additional constructors to $\mathcal{SH}$.

### 4.2.5.1  $\mathcal{SHOIN}$

The $\mathcal{SHOIN}$ description logic, which is the logical underpinning of OWL, can be defined briefly as follows. Let $N_C$, $N_R$, and $N_I$ be countably finite and pairwise disjoint sets of atomic concepts, atomic roles, and individual names. A $\mathcal{SHOIN}$ role expression $\mathbf{R}$ is either an atomic role $N_R \in \mathbf{R}$ or $N_R$-, the inverse of an atomic role $N_R$. A $\mathcal{SHOIN}$ concept expression $\mathbf{C}$ can contain $N_C \in \mathbf{C}$, the $\top$ and $\bot$ concepts, and if $a \in N_I$, then $\{a\}$ is also a concept. If $C$ and $D$ are concepts and $N_R \in \mathbf{R} \cup \mathbf{R}^-$, then $(C \sqcap D)$, $(C \sqcup D)$, $\neg C$, $\exists R.C$, $\forall R.C$, $\geqslant nR$, and $\leqslant nR$, where $n$ is a positive integer, are also concepts.

### 4.2.5.2  $\mathcal{SROIQ}$

The $\mathcal{SROIQ}$ description logic, which is the formal underpinning of OWL 2, can be defined briefly as follows. Assume three countably finite and pairwise disjoint sets of atomic concepts ($N_C$), atomic roles ($N_R$), and individual names ($N_I$), denoted by Internationalized Resource Identifiers (IRIs). In $\mathcal{SROIQ}$, concept expressions can include concept names, concept intersection, concept union, complement, tautology, contradiction, existential and universal quantifiers, qualified at-least and at-most restrictions, local reflexivity, and individual names (see Definition 4.22).

**Definition 4.22 ($\mathcal{SROIQ}$ Concept Expression)** The set of $\mathcal{SROIQ}$ concept expressions is defined as $\mathbf{C} ::= N_C \mid (C \sqcap C) \mid (C \sqcup C) \mid \neg C \mid \top \mid \bot \mid \exists R.C \mid \forall R.C \mid \geqslant nR.C \mid \leqslant nR.C \mid \exists R.Self \mid \{N_I\}$, where $C$ represents concepts, $R$ represents roles, and $n$ is a nonnegative integer.

$\mathcal{SROIQ}$ role expressions support the universal role, atomic roles, and negated atomic roles (see Definition 4.23).

**Definition 4.23 ($\mathcal{SROIQ}$ Role Expression)** The set of $\mathcal{SROIQ}$ role expressions is defined as $\mathbf{R} ::= U \mid N_R \mid N_{R^-}$, where $U$ is the universal role.

Based on the above sets, the $\mathcal{SROIQ}$ axioms are defined in Definition 4.24.

**Definition 4.24 ($\mathcal{SROIQ}$ Axiom)** A $\mathcal{SROIQ}$ axiom is either a general concept inclusion of the form $C \sqsubseteq D$ and $C \equiv D$ for concepts $C$ and $D$, or an individual assertion of the form $C(a)$, $R(a, b)$, $\neg R(a, b)$, $a \approx b$, $a \not\approx b$, where $a, b \in N_I$ denote individual names, $C \in \mathbf{C}$ denotes concept expressions, and $R \in \mathbf{R}$ denotes roles, or a role assertion of the form $R \sqsubseteq S$, $R_1 \circ \ldots \circ R_n \sqsubseteq S$, Asymmetric $(R)$, Reflexive $(R)$, Irreflexive $(R)$, or Disjoint $(R, S)$ for roles $R$, $R_i$, and $S$.

The definition of $\mathcal{SROIQ}$ interpretations is provided in Definition 4.25.

**Definition 4.25 ($\mathcal{SROIQ}$ Interpretation)** A $\mathcal{SROIQ}$ interpretation $\mathcal{I}$ is a pair of the form $\mathcal{I} = (\Delta^{\mathcal{I}}, \cdot^{\mathcal{I}})$, where $\Delta^{\mathcal{I}}$ is a nonempty set (the object domain), and $\cdot^{\mathcal{I}}$ is the interpretation function, which includes the class interpretation function $\cdot^{\mathcal{I}_C}$, which assigns a subset of the object domain to each class, i.e., $\cdot^{\mathcal{I}_C} : A \in N_C \rightarrow A^{\mathcal{I}} \subseteq \Delta^{\mathcal{I}}$, the role interpretation function $\cdot^{\mathcal{I}_R}$, which assigns a set of tuples over the object domain to each role, i.e., $\cdot^{\mathcal{I}_R} : R \in N_R \rightarrow \Delta^{\mathcal{I}} \times \Delta^{\mathcal{I}}$, and the individual interpretation function $\cdot^{\mathcal{I}_I}$, which maps each individual $a \in N_I$ to an element $a^{\mathcal{I}_I} \subseteq \Delta^{\mathcal{I}}$.

## 4.2.6   Spatial Description Logics

In the late 1990s, $\mathcal{ALC}$ was extended with a role-forming predicate operator and a concrete domain for the spatial dimension [117–119]. The first spatial description logic in this family was $\mathcal{ALCRP}^{(\mathcal{D})}$, which defines a concrete domain with a set of predicates representing topological relationships between two spatial objects.[21] In a specific $\mathcal{ALCRP}^{(\mathcal{D})}$ domain model, the roles that represent topological relations can be defined based on properties between concrete objects, which are associated to individuals via specific *features*. The role terms that refer to predicates over a concrete domain are unique to $\mathcal{ALCRP}^{(\mathcal{D})}$ and are extensions of $\mathcal{ALC}^{(\mathcal{D})}$.

In $\mathcal{ALCRP}^{(\mathcal{D})}$, concepts are defined as follows. If $C$ and $D$ are concept terms, $R$ is a role term, $P \in \varphi_{\mathcal{D}}$ is a predicate name with arity $n$, and $u_1, \ldots, u_n$ are feature chains (a composition of features), then $C \sqcap D$, $C \sqcup D$, $\neg C$, $\exists R.C$, $\forall R.C$, and existential restrictions of predicates of the form $\exists u_1, \ldots, u_n.P$ are also concepts.

---

[21]$RP$ stands for "role definitions based on predicates."

Atomic roles in $\mathcal{ALCRP}^{(\mathcal{D})}$ can be role names or feature names. A composition of features of the form $f_1 f_2 \ldots$ is called a *feature chain*. If $P \in \varphi_{\mathcal{D}}$ is a predicate name with arity $n + m$, and $u_1, \ldots, u_n$ and $v_1, \ldots, v_m$ are feature chains, role-forming predicate restrictions of the form $\exists(u_1, \ldots, u_n)(v_1, \ldots, v_m).P$ are complex role terms.

The meaning of these atomic and nonatomic concepts, atomic and complex roles, features and feature chains are determined via the interpretation defined in Definition 4.26.

**Definition 4.26 ($\mathcal{ALCRP}^{(\mathcal{D})}$ Interpretation)** An interpretation $\mathcal{I} = (\Delta^{\mathcal{I}}, \cdot^{\mathcal{I}})$ consists of an abstract domain $\Delta^{\mathcal{I}}$ and an interpretation function $\cdot^{\mathcal{I}}$. The interpretation function maps each concept name $C$ to a subset $C^{\mathcal{I}}$ of $\Delta^{\mathcal{I}}$, each role name $R$ to a subset $R^{\mathcal{I}}$ of $\Delta^{\mathcal{I}} \times \Delta^{\mathcal{I}}$, and each feature name $f$ to a partial function $f^{\mathcal{I}}$ from $\Delta^{\mathcal{I}}$ to $\Delta_{\mathcal{D}} \cup \Delta^{\mathcal{I}}$, where $f^{\mathcal{I}}(a) = x$ will be written as $(a,x) \in f^{\mathcal{I}}$. If $u = f_1 \ldots f_n$ is a feature chain, then $u^{\mathcal{I}}$ denotes the composition $f_1^{\mathcal{I}} \circ \ldots \circ f_n^{\mathcal{I}}$ of the partial functions $f_1^{\mathcal{I}}, \ldots, f_n^{\mathcal{I}}$.

The interpretation function can be extended to arbitrary concept and role terms as follows:

$$(C \sqcap D)^{\mathcal{I}} := C^{\mathcal{I}} \cap D^{\mathcal{I}}$$
$$(C \sqcup D)^{\mathcal{I}} := C^{\mathcal{I}} \cup D^{\mathcal{I}}$$
$$(\neg C)^{\mathcal{I}} := \Delta^{\mathcal{I}} \backslash C^{\mathcal{I}}$$
$$(\exists R.C)^{\mathcal{I}} := \{a \in \Delta^{\mathcal{I}} | \exists b \in \Delta^{\mathcal{I}} : (a,b) \in R^{\mathcal{I}}, b \in C^{\mathcal{I}}\}$$
$$(\forall R.C)^{\mathcal{I}} := \{a \in \Delta^{\mathcal{I}} | \forall b \in \Delta^{\mathcal{I}} : (a,b) \in R^{\mathcal{I}} \to b \in C^{\mathcal{I}}\}$$
$$(\exists u_1, \ldots, u_n.P)^{\mathcal{I}} := \{a \in \Delta^{\mathcal{I}} | \exists x_1, \ldots, x_n \in \Delta_{\mathcal{D}} : (a,x_1) \in u_1^{\mathcal{I}}, \ldots, (a,x_n) \in u_n^{\mathcal{I}}, (x_1, \ldots, x_n) \in P^{\mathcal{D}}\}$$
$$(\exists(u_1, \ldots, u_n)(v_1, \ldots, v_m).P)^{\mathcal{I}} := \{(a,b) \in \Delta^{\mathcal{I}} \times \Delta^{\mathcal{I}} | \exists x_1, \ldots, x_n, y_1, \ldots, y_m \in \Delta_{\mathcal{D}} : (a,x_1) \in u_1^{\mathcal{I}}, \ldots, (a,x_n) \in u_n^{\mathcal{I}}, (b,y_1) \in v_1^{\mathcal{I}}, \ldots, (b,y_m) \in v_m^{\mathcal{I}}, (x_1, \ldots, x_n, y_1, \ldots, y_m) \in P^{\mathcal{D}}\}$$

The expressivity of $\mathcal{ALCRP}^{(\mathcal{D})}$ comes at a price: undecidability. However, decidability can be achieved using a syntax restriction.

The extension of $\mathcal{ALCRP}^{(\mathcal{D})}$ with a ternary role-forming predicate operator and inverse roles is called $\mathcal{ALCRP}^{3(\mathcal{D})}$ [120]. $\mathcal{ALCRP}^{3(\mathcal{D})}$ defines compositional semantics and a *DLR*-style syntax, and supports n-ary spatial relations, such as the direction relation. But syntactic restrictions on concepts and roles are required to ensure decidability of the language.

The spatial relationship between two-dimensional objects in a plane can be described formally not only by their metrical and/or geometric attributes, but also by the qualitative spatial relationships between them, such as by employing the well-known *Region Connection Calculus, RCC8*, which describes regions by their potential relations to each other [121], or the *Egenhofer relations*, which are binary topological relations [122]. RCC8 consists of eight basic relations that are possible between two regions: disconnected (DC), externally connected (EC), equal (EQ), partially overlapping (PO), tangential proper part (TPP), tangential proper part inverse (TPPi), nontangential proper part (NTPP), and nontangential proper part inverse (NTPPi) (see Fig. 4.2).

**Fig. 4.2** Two-dimensional examples for the eight core RCC8 relations

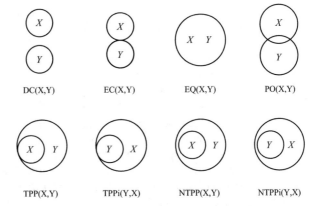

One of the early description logic implementations of RCC8 was $\mathcal{ALC}_{RA\ominus}$, which extended $\mathcal{ALC}$ with composition-based role axioms of the form $S \circ T \sqsubseteq R_1 \sqcup \ldots \sqcup R_n$, $n \geqslant 1$, enforcing $S^{\mathcal{I}} \circ T^{\mathcal{I}} \subseteq R_1^{\mathcal{I}} \cup \ldots \cup R_n^{\mathcal{I}}$ on models $\mathcal{I}$ [123].

The $\mathcal{ALC}(\mathcal{D}_{RCC8})$ description logic was specially designed to implement RCC8 [124]. By using the basic RCC8 relations and their combinations, $\mathcal{ALC}(\mathcal{D}_{RCC8})$ is suitable for qualitative spatial representation and reasoning, where the spatial regions are regular subsets of a topological space.

The RCC8 predicates can also be defined as a set of complex roles using the role-forming predicate-based operator of $\mathcal{ALCRP}^{(\mathcal{D})}$, e.g.,

Disconnected $\doteq \exists(\text{has\_area})(\text{has\_area}).\text{DC}$
Connected $\doteq \exists(\text{has\_area})(\text{has\_area}).\text{C}$

$\mathcal{ALCRP}^{(\mathcal{D})}$ can be used directly for representing spatial relationships between stationary objects in videos as follows:

disconnectedWith $\doteq \exists(\text{concept\_1})(\text{concept\_2}).\text{DC}$
connectedWith $\doteq \exists(\text{concept\_1})(\text{concept\_2}).\text{C}$

Using the above description logic formalism, a topological relationship ontology was proposed for describing moving objects in videos. For moving objects, Na et al. considered line-region topological relationships, where a stationary object depicted in the video is described by a region, and the trajectory of a moving object is represented by a line. Their nine-intersection line-region topological model can be applied to three different object types: area, line, and point. It characterizes the topological relation between two point sets, $A$ and $B$, by the set intersections of $A$'s interior $(A^0)$, boundary $(\partial A)$, and exterior $(A^-)$ with the interior, boundary, and exterior of $B$ (see Eq. 4.1).

$$I(A,B) = \begin{pmatrix} A^0 \cap B^0 & A^0 \cap \partial B & A^0 \cap B^- \\ \partial A \cap B^0 & \partial A \cap \partial B & \partial A \cap B^- \\ A^- \cap B^0 & A^- \cap \partial B & A^- \cap B^- \end{pmatrix} \qquad (4.1)$$

Between a simple line (one-dimensional, nonbranching, without self-intersections) and a region (two-dimensional, simply connected, no holes) embedded in $R^2$, there are 19 different line-region relationships with the nine-intersection model. Based on the deformations that would cause a topological relationship to change because of the boundary or interior of the line is pulled or pushed, these relationships can be arranged according to their topological neighborhoods, resulting in a neighborhood graph. This can be utilized to describe moving objects in video by clustering motion verbs for direction and movement, as, for example, cross, go through, arrive, enter, exit, depart, go away, and return.

$\mathcal{ALCI}_{RCC}$ is a family of spatial description logics that extends the standard $\mathcal{ALCI}$ description logic with the Region Connection Calculus to be sufficient for qualitative spatial reasoning tasks [125]. Members of the $\mathcal{ALCI}_{RCC}$ family include $\mathcal{ALCI}_{RCC1}$, $\mathcal{ALCI}_{RCC2}$, $\mathcal{ALCI}_{RCC3}$, $\mathcal{ALCI}_{RCC5}$, and $\mathcal{ALCI}_{RCC8}$, each of which provides RCC relationships on various levels of granularity, where the set of roles $N_R$ corresponds to different sets of RCC relationships, namely, $N_{R\,\mathcal{ALCI}_{RCC8}} = \{$DC, EC, PO, EQ, TPP, TPPi, NTPP, NTPPi$\}$, $N_{R\,\mathcal{ALCI}_{RCC5}} = \{$DR, PO, EQ, PP, PPi$\}$, $N_{R\,\mathcal{ALCI}_{RCC3}} = \{$DR, ONE, EQ$\}$, $N_{R\,\mathcal{ALCI}_{RCC2}} = \{$DR, O$\}$, and $N_{R\,\mathcal{ALCI}_{RCC1}} = \{$SR$\}$.[22]

$\mathcal{ALC}$(F) is a spatial description logic specially designed for spatial reasoning over images [126]. $\mathcal{ALC}$(F) extends $\mathcal{ALC}$ with concrete domains to integrate quantitative and qualitative characteristics of real-world objects in a conceptual domain featuring mathematical morphological operators as predicates. This enables useful concrete representations of spatial concepts, quantitative and qualitative spatial reasoning, and handling imprecision and uncertainty of spatial knowledge via fuzzy representation of spatial relations.

$\mathcal{ALC}$(CDC) is the extension of $\mathcal{ALC}$ with the *Cardinal Direction Calculus* (CDC), which can be defined briefly as follows. Given a bounded region $b$ in the real plane, the plane is partitioned into nine tiles by extending the four edges of the minimum bounding rectangle (MBR) of $b$, $\mathcal{M}(b)$, as shown in Fig. 4.3 [127].

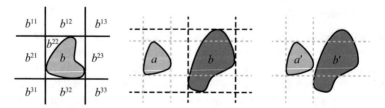

**Fig. 4.3** Cardinal direction calculus

---

[22]The RCC5 relationships are obtained from the set of RCC8 relationships by collapsing DC and EC into DR ("Discrete"), TPP and NTPP into PP ("Proper Part"), as well as TPPi and NTPPi into PPi ("Proper Part Inverse"). In RCC3, {PP, PPi, PO} are collapsed into ONE ("Overlapping but Not Equal"), and in RCC2 {ONE, EQ} into O ("Overlapping"). RCC1 has only one relationship called SR ("Spatially Related").

For a primary region $a$ and a reference region $b$, the CDC relation of $a$ to $b$, $\delta_{ab}$, is encoded in a $3 \times 3$ Boolean matrix $(d_{ij})_{1 \leqslant i,j \leqslant 3}$, where $d_{ij} = 1$ iff $a^{\circ} \cap b^{ij} \neq \emptyset$, where $a^{\circ}$ is the interior of $a$. Since a CDC relation can be any Boolean matrix except the zero Boolean matrix, there are 511 basic relations in CDC.[23]

The combination of $\mathcal{ALC}(\text{RCC})$ and $\mathcal{ALC}(\text{CDC})$ is called $\mathcal{ALC}(\text{CDRCC})$.

### 4.2.7 Temporal Description Logics

Temporal description logics formally represent, and enable reasoning about, time-dependent concepts, actions, and video events. The main approaches to employ standard temporal logic operators,[24] such as the unary operators ○ (*at the next state*), ◇ (*in the future/eventually*), and □ (*globally*)[25] and the binary operator U[26] (*until*) (see Fig. 4.4), with general-purpose description logics aimed at representing terminological knowledge of a domain include the following:

1. Add temporal operators as additional concept constructors
2. Apply the temporal operators to the TBox, the ABox, and roles
3. Combine the above two approaches

Such combinations are based on two-dimensional semantics, where one dimension represents time using point-based or interval-based temporal logic and the other the knowledge domain using a standard description logic.

In the combinations of description logics and point-based temporal propositional logics, the description logic component represents knowledge about the process state, and the temporal component describes the sequence of states, i.e., the behavior of processes over time (see Definition 4.27).

**Definition 4.27 (Temporal Interpretation)** A temporal interpretation $\mathcal{I} = (\Delta^{\mathcal{I}}, \cdot^{\mathcal{I}})$ consists of a nonempty domain $\Delta^{\mathcal{I}}$ and a function $\cdot^{\mathcal{I}}$ that maps every concept name $A \in N_C$ to a subset $A^{\mathcal{I}} \subseteq \mathbb{N} \times \Delta^{\mathcal{I}}$, every role name $R \in N_R$ to a subset

---

[23]Calculated as $2^9 - 1$

[24]This results in additional annotations beyond the standard description logic nomenclature. Note that the use of the CMSY10 font for temporal description logic names is inconsistent in the literature. Sometimes they are used for the entire DL name regardless that the temporal part, unlike the DL part, does not correspond to mathematical constructors. Also, some authors write the DL part as a subscript while others write a dash between the temporal and the DL parts. This book applies a consistent, correct, and unified annotation throughout.

[25]Read as circle, diamond, and box (or square) operator, respectively.

[26]The symbolic annotation of the *until* operator is written in the CMSY10 font in the temporal logic literature, i.e., $\mathcal{U}$ (see Fig. 4.4). Since this could be confused with the $\mathcal{U}$ used in description logic names for concept union, the textual annotation of the *until* operator is used throughout this book instead, which is written as U.

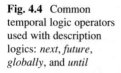

**Fig. 4.4** Common temporal logic operators used with description logics: *next*, *future*, *globally*, and *until*

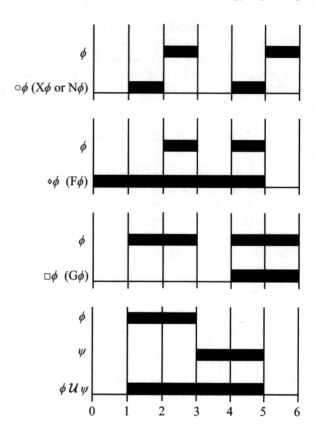

$R^{\mathcal{I}} \subseteq \mathbb{N} \times \Delta^{\mathcal{I}} \times \Delta^{\mathcal{I}}$, and every individual name $a \in N_R$ to an element $a^{\mathcal{I}} \in \Delta^{\mathcal{I}}$, where the elements of $\mathbb{N}$ represent time points ordered by $<$.

Logics combined with conventional description logics to form temporal description logics include, but are not limited to, Prior's *Tense Logic (TL)*, *Linear Temporal Logic (LTL)*, *Computational Tree Logic (CTL)*, and the time interval logic of Halpern and Shoham [128]. Among others, temporal description logics are suitable for the formal representation of video actions and employ datatypes for time points and time intervals [129].

The basic temporal description logic *TL-F* is composed of the temporal language *TL*, which expresses interval temporal networks, and a nontemporal *feature description logic F*. More expressive temporal description logics include *TLU-FU*, which adds disjunction to both the temporal and the nontemporal sides of *TL-F*, and *TL-ALCF*, which extends the nontemporal side of *TL-F* with roles and full propositional calculus [130]. The subsumption problem for these languages is decidable [131]. Since adding temporal operators to description logics might result in undecidability, such as recursively nonenumerable formalisms, only the propositional fragment of temporal logics is researched and implemented.

### 4.2.7.1  Temporal Extensions of $\mathcal{ALC}$

$T$-$\mathcal{ALC}$ is the temporal extension of the core description logic $\mathcal{ALC}$ with time intervals (see Definition 4.28), which are suitable for, among others, reasoning over actions of video events [132].

**Definition 4.28 ($T$-$\mathcal{ALC}$ Time Interval)**  A $T$-$\mathcal{ALC}$ time interval $T$ is an ordered pair $[t_1, t_2]$, $(t_1 \leqslant t_2).t_1 = \text{start}(T)$ is the start time of $T$, $t_2 = \text{end}(T)$ is the end time of $T$, where $t_1$ and $t_2$ are time points, and $T$ represents a time point when $t_1 = t_2$.

A knowledge base with temporal information, $\mathcal{K}_{T\text{-}\mathcal{ALC}}$, consists of an $\mathcal{ALC}$ TBox $\mathcal{T}$ and a finite set of assertional axioms complemented by temporal information, $\mathcal{A}_T$ (see Definition 4.29).

**Definition 4.29 ($T$-$\mathcal{ALC}$ ABox)**  An ABox which contains concept assertions of the form $C(a[t_1,t_2])$ or $C(a^t)$ and role assertions of the form $R(a,b^{[t1,t2]})$ or $R((a,b)^t)$, where $a, b \in N_I, C \in N_C, R \in N_R$, and $t_1, t_2$ are time points, is called a $T$-$\mathcal{ALC}$ ABox, and is usually denoted by $\mathcal{A}_T$.

The interpretation of $T$-$\mathcal{ALC}$ ABox axioms is defined in Definition 4.30.

**Definition 4.30 (Interpretation in $T$-$\mathcal{ALC}$ ABox)**  An interpretation of $T$-$\mathcal{ALC}$ in ABox $\mathcal{A}_T$ is of the form $\mathcal{I}(t) = (\Delta^{\mathcal{I}(t)}, \cdot^{\mathcal{I}(t)})$. $\Delta^{\mathcal{I}(t)}$ is the domain of $\mathcal{I}(t)$ which is nonempty set of individuals at $t$ and $\cdot^{\mathcal{I}(t)}$ is an interpretation function at $t$ such that for each concept $C(a)$ if there exists $C(a^t)$, then $t, a^{\mathcal{I}(t)} \in C^{\mathcal{I}(t)}, C \subseteq \Delta^{\mathcal{I}(t)}$, if there exists $C$ $(a[t_1,t_2])$ and $t \in [t_1,t_2]$, then at $t$, $a^{\mathcal{I}(t)} \in C^{\mathcal{I}(t)}, C^{\mathcal{I}(t)} \subseteq \Delta^{\mathcal{I}(t)}$; for each role $R(a,b)$, if there exists $R(a,b)^t$, then at $t$, $(a,b)^{\mathcal{I}(t)} \in R^{\mathcal{I}(t)}, R^{\mathcal{I}(t)} \subseteq \Delta^{\mathcal{I}(t)} \times \Delta^{\mathcal{I}(t)}$, if there exists $R$ $(a,b[t_1,t_2])$ and $t \in [t_1,t_2]$, then at $t$, $(a,b)^{\mathcal{I}(t)} \in R^{\mathcal{I}(t)}, R^{\mathcal{I}(t)} \subseteq \Delta^{\mathcal{I}(t)} \times \Delta^{\mathcal{I}(t)}$.

$TL$-$\mathcal{ALC}$ employs the temporal operators $\circ$ (*next*) and $\mathsf{U}$ (*until*) applied to concepts and formulae, and interpreted over the natural numbers with expanding $\mathcal{ALC}$ domains [133]. The syntax and semantics of $TL$-$\mathcal{ALC}$ are summarized in Table 4.12.

The $TL$-$\mathcal{ALC}$ model maps natural numbers to $\mathcal{ALC}$ models, as described in Definition 4.31.

**Definition 4.31 ($TL$-$\mathcal{ALC}$ Model)**  A $TL$-$\mathcal{ALC}$ model is a triple of the form $\mathfrak{M} = (\mathbb{N}, <, \mathcal{I})$, where $(\mathbb{N}, <)$ is the set of natural numbers in strict order, and $\mathcal{I}$ is the interpretation function associating with each $n \in \mathbb{N}$ some $\mathcal{ALC}$ model $\mathcal{I}(n) = (\Delta^n, R_0^{\mathcal{I},n}, \ldots, C_0^{\mathcal{I},n}, \ldots, a_0^{\mathcal{I},n}, \ldots)$, in which $R_i^{\mathcal{I},n}$ are binary relations on $\Delta^n$, $C_0^{\mathcal{I},n}$ are subsets of $\Delta^n$, $a_0^{\mathcal{I},n} \in \Delta^{\mathcal{I},n}$ such that $a_i^{\mathcal{I},n} = a_i^{\mathcal{I},m}$ for every $n,m \in \mathbb{N}$, and $\Delta^n \subseteq \Delta^m$ for every $n,m \in \mathbb{N}$ with $n < m$. A $TL_{\mathcal{ALC}}$ formula $\varphi$ is called satisfiable if and only if there exists a model $\mathfrak{M}$ and a moment $n \in \mathbb{N}$ such that $\mathfrak{M}, n \vDash \varphi$.

The temporal description logic $TL$-$\mathcal{ALCF}$ is composed of the interval-based temporal logic $TL$ and the nontemporal description logic $\mathcal{ALCF}$, and was designed for domains that have objects with properties that vary over time. The $TL$ part represents temporal constraint networks based on Allen's temporal relations [134] (see Fig. 4.5) and maps $\mathcal{ALCF}$ concept expressions to time intervals, which enables reasoning over actions [135].

**Table 4.12** $TL\text{-}\mathcal{ALC}$ syntax and semantics

|  | Syntax | Semantics |
|---|---|---|
| Objects | $a_0, a_1, \ldots$ | $a_0^{\mathcal{I},n} \in \Delta^{\mathcal{I},n}, \ldots$ |
| Roles | $R_0, R_1, \ldots$ | $R_0^{\mathcal{I},n} \subseteq \Delta^n \times \Delta^n, \ldots$ |
| Concepts | $C_0, C_1, \ldots$ | $C_0^{\mathcal{I},n} \subseteq \Delta^n, \ldots$ |
|  | $\top$ | $\top^{\mathcal{I},n} = \Delta^n$ |
|  | $C \sqcap D$ | $(C \sqcap D)^{\mathcal{I},n} = C^{\mathcal{I},n} \cap D^{\mathcal{I},n}$ |
|  | $\neg C$ | $(\neg C)^{\mathcal{I},n} = \Delta^n \backslash C^{\mathcal{I},n}$ |
|  | $\exists R_i.C$ | $(\exists R_i.C)^{\mathcal{I},n} = \{ d \in \Delta^n | \exists d' \in C^{\mathcal{I},n} : dR_i^{\mathcal{I},n}d' \}$ |
|  | $\circ C$ | $(\circ C)^{\mathcal{I},n} = \{ d \in \Delta^n | d \in C^{\mathcal{I},n+1} \}$ |
|  | $C \mathcal{U} D$ | $(C \mathcal{U} D)^{\mathcal{I},n} = \{ d \in \Delta^n | \exists m \geqslant n (d \in D^{\mathcal{I},m}$ and $\forall k (n \leqslant k < m \Rightarrow d \in C^{\mathcal{I},k})) \}$ |
| Formulae | $C = D$ | $\mathfrak{M}, n \vDash C = D$ iff $C^{\mathcal{I},n} = D^{\mathcal{I},n}$ |
|  | $C(a)$ | $\mathfrak{M}, n \vDash C(a)$ iff $a^{\mathcal{I},n} \in C^{\mathcal{I},n}$ |
|  | $\top$ | $\mathfrak{M}, n \vDash \top$ |
|  | $\neg \varphi$ | $\mathfrak{M}, n \vDash \neg \varphi$ iff $\mathfrak{M}, n \nvDash \varphi$ |
|  | $\varphi \wedge \psi$ | $\mathfrak{M}, n \vDash \varphi \wedge \psi$ iff $\mathfrak{M}, n \vDash \varphi$ and $\mathfrak{M}, n \vDash \psi$ |
|  | $\circ \varphi$ | $\mathfrak{M}, n \vDash \circ \varphi$ iff $\mathfrak{M}, n + 1 \vDash \varphi$ |
|  | $\varphi \mathcal{U} \psi$ | $\mathfrak{M}, n \vDash \varphi \mathcal{U} \psi$ iff there is some $m \geqslant n$ with $\mathfrak{M}, m \vDash \varphi$ such that for all $k$ with $n \leqslant k < m, \mathfrak{M}, k \vDash \varphi$ |

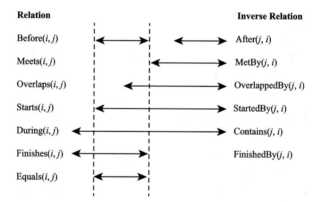

**Fig. 4.5** Allen's temporal operators

The syntax rules of the *TL* part of *TL-$\mathcal{ALCF}$* include nontemporal concepts, conjunction, qualifiers that specify which concepts are TRUE at intervals denoted by temporal variables, substitutive qualifiers, temporal qualifiers, temporal constraints, disjunction, Allen's relations, and temporal variables. The $\mathcal{ALCF}$ syntax rules cover atomic concepts, complement, conjunction, disjunction, the universal quantifier, the existential quantifier, selection, agreement, disagreement, undefinedness, atomic features, atomic parametric features (features independent from time), and paths (see Table 4.13).

**Table 4.13** *TL-ALCF* syntax rules

| TL part |
|---|
| $C,D \rightarrow E \mid C \sqcap D \mid C@X \mid C[Y]@X \mid \diamond(\overline{X}), \overline{T_C}.C$ |
| $T_C \rightarrow (X(r)Y) \mid (X(r) \sharp) \mid (\sharp(r)Y)$ |
| $\overline{T_C} \rightarrow T_C \mid T_C\overline{T_C}$ |
| $r,s \rightarrow r,s \mid b \mid m \mid d \mid o \mid s \mid f \mid = \mid a \mid mi \mid di \mid oi \mid si \mid fi$ |
| $X,Y \rightarrow x \mid y \mid z \mid \ldots$ |
| $\overline{X} \rightarrow X \mid X\overline{X}$ |

| ALCF part |
|---|
| $E,F \rightarrow A \mid \neg E \mid E \sqcap F \mid E \sqcup F \mid \forall R.E \mid \exists R.E \mid p : E \mid p{\downarrow}q \mid p{\uparrow}q \mid p{\uparrow}$ |
| $p,q \rightarrow f \mid \star g \mid p \bigcirc q$ |

The two-dimensional semantics of *TL-ALCF* is defined by the temporal constraint networks that occur inside the $\diamond$ operator (see Table 4.14) and the interpretation of *TL-ALCF* concepts (see Definition 4.32).

**Definition 4.32 (TL-ALCF Interpretation)** Assume the temporal structure $\mathcal{T} = (P, <)$, where $P$ is a set of time points and $<$ is a linear, unbounded, and dense order on $P$. The interval set of structure $\mathcal{T}$ is defined as the set $\mathcal{T}_<^\star$ of all closed proper intervals of the form $[u, v] \doteq \{x \in P \mid u \leqslant x \leqslant v, u \neq v\}$ in $\mathcal{T}$. A *TL-ALCF* interpretation $\mathcal{I} \doteq \langle \mathcal{T}_<^\star, \Delta^{\mathcal{T}}, \cdot^{\mathcal{I}} \rangle$ consists of the interval set of the selected temporal structure $\mathcal{T}$, the domain of $\mathcal{I}$ $\Delta^{\mathcal{I}}$, and the interpretation function $\cdot^{\mathcal{I}}$, which provides the meaning for atomic concepts, roles, features, and parametric features as follows: $A^{\mathcal{I}} \subseteq \mathcal{T}_<^\star \times \Delta^{\mathcal{I}}, R^{\mathcal{I}} \subseteq \mathcal{T}_<^\star \times \Delta^{\mathcal{I}} \times \Delta^{\mathcal{I}}, f^{\mathcal{I}} : \left( \mathcal{T}_<^\star \times \Delta^{\mathcal{I}} \right) \overset{\text{partial}}{\mapsto} \Delta^{\mathcal{I}}$, and $\star g^{\mathcal{I}} : \Delta^{\mathcal{I}} \overset{\text{partial}}{\mapsto} \Delta^{\mathcal{I}}$.

The temporal part of *ALCF*(A) is provided by the temporal concept constructor $E$, $F \rightarrow \exists p_1, p_2.r$, where $r$ is an Allen relation and $p_1, \ldots, p_n$ are temporal paths, i.e., sequences $\gamma_1 \circ \ldots \circ \gamma_k \circ h$, where $\gamma_1, \ldots, \gamma_k$ are features and $h$ is a temporal feature.

The definition of interpretations in *ALCF*(A) is provided by Definition 4.33.

**Definition 4.33 (ALCF(A) Interpretation)** An *ALCF*(A) interpretation $\mathcal{I} = (\Delta^{\mathcal{I}}, \cdot^{\mathcal{I}})$ consists of a set $\Delta^{\mathcal{I}}$ and an interpretation function $\cdot^{\mathcal{I}}$ which assigns a meaning to concept names, role names, features, and temporal features as follows:

$$A^{\mathcal{I}} \subseteq \Delta^{\mathcal{I}}$$
$$R^{\mathcal{I}} \subseteq \Delta^{\mathcal{I}} \times \Delta^{\mathcal{I}}$$
$$f^{\mathcal{I}} : \Delta^{\mathcal{I}} \overset{\text{partial}}{\mapsto} \Delta^{\mathcal{I}}$$
$$h^{\mathcal{I}} : \Delta^{\mathcal{I}} \overset{\text{partial}}{\mapsto} \mathcal{T}_<^*$$

The semantics of the temporal concept constructor is defined as $(\exists p_1, p_2.r)^{\mathcal{I}} = \{a \in \Delta^{\mathcal{I}} \mid \exists t_1, t_2 \in \mathcal{T}_<^* : (a, t_1) \in p_1^{\mathcal{I}} \wedge (a, t_2) \in p_2^{\mathcal{I}} \wedge (t_1, t_2) \in r^\varepsilon \}$.

The extension of *ALC* with Linear Temporal Logic (LTL) is called *ALC*-LTL (or LTL$_{ALC}$), which is suitable for specifying wanted and unwanted properties of dynamic systems [136].

**Table 4.14** *TL-$\mathcal{ALCF}$ semantics*

$$(s)^\varepsilon = \left\{ ([u,v],[u_1,v_1]) \in \mathcal{T}^\star_< \times \mathcal{T}^\star_< | u = u_1 \wedge v < v_1 \right\}$$
... (analogously for all the other Allen relations)
$$(r,s)^\varepsilon = r^\varepsilon \cup s^\varepsilon$$
$$\langle \overline{X}, \overline{T_C} \rangle^\varepsilon = \left\{ \mathcal{V} : \overline{X} \mapsto \mathcal{T}^\star_< | \forall (X\, r\, Y) \in \overline{T_C}.(\mathcal{V}(X), \mathcal{V}(Y)) \in r^\varepsilon \right\}$$
$$A^\mathcal{I}_{\mathcal{V},t,\mathcal{H}} = \left\{ a \in \Delta^\mathcal{I} | (t,a) \in A^\mathcal{I} \right\}$$
$$(\neg C)^\mathcal{I}_{\mathcal{V},t,\mathcal{H}} = \Delta^\mathcal{I} \setminus C^\mathcal{I}_{\mathcal{V},t,\mathcal{H}}$$
$$(C \sqcap D)^\mathcal{I}_{\mathcal{V},t,\mathcal{H}} = C^\mathcal{I}_{\mathcal{V},t,\mathcal{H}} \cap D^\mathcal{I}_{\mathcal{V},t,\mathcal{H}}$$
$$(C \sqcup D)^\mathcal{I}_{\mathcal{V},t,\mathcal{H}} = C^\mathcal{I}_{\mathcal{V},t,\mathcal{H}} \cup D^\mathcal{I}_{\mathcal{V},t,\mathcal{H}}$$
$$(\forall R.C)^\mathcal{I}_{\mathcal{V},t,\mathcal{H}} = \left\{ a \in \Delta^\mathcal{I} | \forall b.(a,b) \in R^\mathcal{I}_t \Rightarrow b \in C^\mathcal{I}_{\mathcal{V},t,\mathcal{H}} \right\}$$
$$(\exists R.C)^\mathcal{I}_{\mathcal{V},t,\mathcal{H}} = \left\{ a \in \Delta^\mathcal{I} | \exists b.(a,b) \in R^\mathcal{I}_t \wedge b \in C^\mathcal{I}_{\mathcal{V},t,\mathcal{H}} \right\}$$
$$(p : C)^\mathcal{I}_{\mathcal{V},t,\mathcal{H}} = \left\{ a \in \mathrm{dom}p^\mathcal{I}_t | p^\mathcal{I}_t(a) \in C^\mathcal{I}_{\mathcal{V},t,\mathcal{H}} \right\}$$
$$(p \downarrow q)^\mathcal{I}_{\mathcal{V},t,\mathcal{H}} = \left\{ a \in \mathrm{dom}p^\mathcal{I}_t \cap \mathrm{dom}q^\mathcal{I}_t | p^\mathcal{I}_t(a) = q^\mathcal{I}_t(a) \right\}$$
$$(p \uparrow q)^\mathcal{I}_{\mathcal{V},t,\mathcal{H}} = \left\{ a \in \mathrm{dom}p^\mathcal{I}_t \cap \mathrm{dom}q^\mathcal{I}_t | p^\mathcal{I}_t(a) \neq q^\mathcal{I}_t(a) \right\}$$
$$(p \uparrow)^\mathcal{I}_{\mathcal{V},t,\mathcal{H}} = \Delta^\mathcal{I} \setminus \mathrm{dom}p^\mathcal{I}_t$$
$$(C@X)^\mathcal{I}_{\mathcal{V},t,\mathcal{H}} = C^\mathcal{I}_{\mathcal{V},\mathcal{V}(X),\mathcal{H}}$$
$$(C[Y]@X)^\mathcal{I}_{\mathcal{V},t,\mathcal{H}} = C^\mathcal{I}_{\mathcal{V},t,\mathcal{H} \cup \{Y \mapsto \mathcal{V}(X)\}}$$
$$\left( \diamond (\overline{X}), \overline{T_C}.C \right)^\mathcal{I}_{\mathcal{V},t,\mathcal{H}} = \left\{ a \in \Delta^\mathcal{I} | \exists \mathcal{W}.\mathcal{W} \in \langle \overline{X}, \overline{T_C} \rangle^\varepsilon_{\mathcal{H} \cup \{\# \mapsto t\}} \wedge a \in C^\mathcal{I}_{\mathcal{W},t,0} \right\}$$
$$R^\mathcal{I}_t = \left\{ (a,b) \in \Delta^\mathcal{I} \times \Delta^\mathcal{I}, | (t,a,b) \in R^\mathcal{I} \right\}$$
$$f^\mathcal{I}_t(a) = b \text{ iff } f^\mathcal{I}(t,a) = b$$
$$(\gamma \circ q)^\mathcal{I}_t(a) = b \text{ iff } q^\mathcal{I}_t \left( \gamma^\mathcal{I}_t(a) \right) = b$$
$$\star g^\mathcal{I}_t = \star g^\mathcal{I}$$

*A* denotes atomic concepts, *C* and *D* denote temporal *TL-$\mathcal{ALCF}$* concepts, *E* and *F* denote (nontemporal) $\mathcal{ALCF}$ concepts, *R* denotes roles, *f* denotes nonparametric features, $\star g$ denotes parametric features, *p* and *q* denote paths, i.e., finite sequences of the form $g_1, \ldots, g_k$, where each $g_i$ is a feature or a parametric feature. *X* and *Y* denote temporal variables. The temporal interval relations are denoted by *r* and *s* using Allen's notation: b (*before*), m (*meets*), d (*during*), o (*overlaps*), s (*starts*), f (*finishes*), = (*equal*), a (*after*), mi (*met by*), di (*contains*), oi (*overlapped by*), si (*started by*), and fi (*finished by*). The temporal existential quantifier $\diamond$ introduces interval variables related to each other and to variable $\sharp$ (*now*, a special temporal variable, which serves as a reference) according to a set of temporal constraints. *C@X* enables the evaluation of concept *C* at an interval *X* different from the current one by temporally qualifying it at *X*. $\star$ distinguishes parametric and nonparametric features and is not an operator

Each $\mathcal{ALC}$ axiom corresponds to an $\mathcal{ALC}$-LTL formula with past operators. Also, if $\varphi$ and $\psi$ are $\mathcal{ALC}$-LTL formulae, then $\neg\varphi$, $\circ\varphi$, $\circ^-\varphi$, $\varphi \wedge \psi$, and $\varphi\ \mathsf{U}\ \psi$ are also $\mathcal{ALC}$-LTL formulae.

Structures in $\mathcal{ALC}$-LTL can be defined as restrictions of a sequence of $\mathcal{ALC}$ interpretations (see Definition 4.34).

**Definition 4.34 ($\mathcal{ALC}$-LTL Structure)** An $\mathcal{ALC}$-LTL structure is a sequence $\mathcal{I} = (\mathcal{I}_i)_{i \geqslant 0}$ of $\mathcal{ALC}$ interpretations $\mathcal{I}_i = (\Delta, {}^{\mathcal{I}_i})$ such that $a^{\mathcal{I}_i} = a^{\mathcal{I}_j}$ for all $a \in N_I$ and all $i, j \geqslant 0$. Given an $\mathcal{ALC}$-LTL formula $\varphi$, an $\mathcal{ALC}$-LTL structure $\mathcal{I} = (\mathcal{I}_i)_{i \geqslant 0}$ and a time point $i \geqslant 0$, validity of $\varphi$ in $\mathcal{I}$ at time $i$ is defined inductively as follows:

$\mathcal{I},i \vDash \alpha$ iff $\mathcal{I}_i \vDash \alpha$ for $\mathcal{ALC}$ axioms $\alpha$

$\mathcal{I},i \vDash \varphi \wedge \psi$ iff $\mathcal{I},i \vDash \varphi$ and $\mathcal{I},i \vDash \psi$

$\mathcal{I},i \vDash \neg\varphi$ iff not $\mathcal{I},i \vDash \varphi$

$\mathcal{I},i \vDash \circ\varphi$ iff $\mathcal{I},i+1 \vDash \varphi$

$\mathcal{I},i \vDash \circ^-\varphi$ iff $i > 0$ and $\mathcal{I},i-1 \vDash \varphi$

$\mathcal{I},i \vDash \varphi \cup \psi$ iff there is some $k \geqslant i$ such that $\mathcal{I},k \vDash \psi$ and $\mathcal{I},j \vDash \varphi$ for all $j, i \leqslant j \leqslant k$

Based on the $\mathcal{ALC}$-LTL structures, the $\mathcal{ALC}$-LTL models can be defined (see Definition 4.35).

**Definition 4.35 ($\mathcal{ALC}$-LTL Model)** The $\mathcal{ALC}$-LTL structure $\mathcal{I}$ is called a model of the $\mathcal{ALC}$-LTL formula $\varphi$ if $\mathcal{I},0 \vDash \varphi$. The $\mathcal{ALC}$-LTL formula $\varphi$ is satisfiable if it has a model.

The combination and modification of the description logic $\mathcal{ALC}$ and Prior's Tense Logic led to the $X\mathcal{ALC}$ and $B\mathcal{ALC}_l$ temporal description logics [137]. $X\mathcal{ALC}$ has a *next* operator, and $B\mathcal{ALC}_l$ has restricted versions of the *next*, *anytime*, and *sometime* operators, where the time domain is bounded by a positive integer $l$.

The definition of $X\mathcal{ALC}$ concepts and interpretation are given in Definitions 4.36 and 4.37, respectively.

**Definition 4.36 ($X\mathcal{ALC}$ Concept)** $X\mathcal{ALC}$ concepts are defined as $\mathbf{C} :: = A \mid \neg C \mid XC \mid C \sqcap D \mid C \sqcup D \mid \forall R.C \mid \exists R.C$.

**Definition 4.37 ($X\mathcal{ALC}$ Temporal Interpretation)** In $X\mathcal{ALC}$, a temporal interpretation $T\mathcal{I}$ is a structure $\left\langle \Delta^{T\mathcal{I}}, \left\{ \mathcal{I}^i \right\}_{i \in \omega} \right\rangle$, where $\Delta^{T\mathcal{I}}$ is a nonempty set and each $\mathcal{I}^i$ $(i \in \omega)$ is an interpretation function, which assigns to every atomic concept $A$ a set $A^{\mathcal{I}^i} \subseteq \Delta^{T\mathcal{I}}$, for any atomic role $R$ and any $i,j \in \omega$, $R^{\mathcal{I}^i} = R^{\mathcal{I}^j}$. The interpretation function is extended to concepts by the following inductive definitions:

$$(XC)^{\mathcal{I}^i} := C^{\mathcal{I}^{i+1}}$$

$$(\neg C)^{\mathcal{I}^i} := \Delta^{T\mathcal{I}} \backslash C^{\mathcal{I}^i}$$

$$(C \sqcap D)^{\mathcal{I}^i} := C^{\mathcal{I}^i} \cap D^{\mathcal{I}^i}$$

$$(C \sqcup D)^{\mathcal{I}^i} := C^{\mathcal{I}^i} \cup D^{\mathcal{I}^i}$$

$$(\forall R.C)^{\mathcal{I}^i} := \left\{ a \in \Delta^{T\mathcal{I}} \mid \forall b \left[ (a,b) \in R^{\mathcal{I}^i} \Rightarrow b \Rightarrow C^{\mathcal{I}^i} \right] \right\}$$

$$(\exists R.C)^{\mathcal{I}^i} := \left\{ a \in \Delta^{T\mathcal{I}} \mid \exists b \left[ (a,b) \in R^{\mathcal{I}^i} \wedge b \in C^{\mathcal{I}^i} \right] \right\}$$

For any $i \in \omega$, an expression $\mathcal{I}^i \vDash C$ is defined as $C^{\mathcal{I}^i} \neq \emptyset$. A temporal interpretation $T\mathcal{I} := \left\langle \Delta^{T\mathcal{I}}, \left\{ \mathcal{I}^i \right\}_{i \in \omega} \right\rangle$ is a model of a concept $C$ if $\mathcal{I}^0 \vDash C$. A concept $C$ is satisfiable in $X\mathcal{ALC}$ if there is a temporal interpretation $T\mathcal{I}$ such that $T\mathcal{I} \vDash C$.

The definitions of $B\mathcal{ALC}_l$ concepts and interpretation are given in Definitions 4.38 and 4.39, respectively.

**Definition 4.38 ($B\mathcal{ALC}_l$ Concepts)** $B\mathcal{ALC}_l$ concepts are defined as $C ::= A \mid \neg C \mid XC \mid GC \mid FC \mid C \sqcap D \mid C \sqcup D \mid \forall R.C \mid \exists R.C$

**Definition 4.39 ($B\mathcal{ALC}_l$ Interpretation)** A bounded time interpretation $BI$ is a temporal structure $\left\langle \Delta BT, \left\{ \mathcal{I}^i \right\}_{i \in \omega} \right\rangle$, where each $\mathcal{I}^i (i \in \omega)$ is an interpretation function, which assigns to every atomic concept $A$ a set $A^{\mathcal{I}^i} \subseteq A^{\mathcal{TI}}$; for any atomic role $R$ and any $i,j \in \omega$, $R^{\mathcal{I}^i} = R^{\mathcal{I}^j}$; for any $i \leqslant l-1, (XC)^{\mathcal{I}^i} := C^{\mathcal{I}^{i+1}}$; for any $i \geqslant l, (XC)^{\mathcal{I}^i} := C^{\mathcal{I}^l}$; and for any $m \in \omega, (XC)^{\mathcal{I}^{l+m}} := C^{\mathcal{I}^l}, (GC)^{\mathcal{I}^i} := C^{\mathcal{I}^i} \cap C^{\mathcal{I}^{i+1}} \cap \ldots \cap C^{\mathcal{I}^{i+l}}$, and $(FC)^{\mathcal{I}^i} := C^{\mathcal{I}^i} \cup C^{\mathcal{I}^{i+1}} \cup \ldots \cup C^{\mathcal{I}^{i+l}}$.

The $\mathcal{EL}$ family of description logics has also been combined with Computational Tree Logic (CTL) fragments, including $E\diamond$ (*possibly eventually*) and $E\circ$ (*possibly at the next state*), resulting in the temporal description logics $\text{CTL}_{\mathcal{EL}}^{E\diamond}$, $\text{CTL}_{\mathcal{EL}}^{E\circ}$, and $\text{CTL}_{\mathcal{EL}}^{E\circ,E\diamond}$ [138]. Using only a restricted set of temporal operators with $\mathcal{EL}$ yields to a relatively low computational complexity compared to other, more expressive temporal description logics.

### 4.2.7.2  $DLR_{US}$

$DLR_{US}$ is the extension of the description logic $DLR$, whose basic elements are concepts and $n$-ary relations, with the temporal operators *until* and *since* (hence the $_{US}$ subscript), making it suitable for creating temporal conceptual models (timestamping and evolution constraints) [139]. It is similar to the $TL\text{-}\mathcal{ALCF}$ temporal description logic; however, $DLR_{US}$ provides a point-based temporal structure, not an interval-based one.

### 4.2.7.3  $TL\text{-}\mathcal{SI}$

$TL\text{-}\mathcal{SI}$ consists of the temporal logic $TL$ and the description logic $\mathcal{SI}$ [140]. The temporal constraint in $TL\text{-}\mathcal{SI}$ is defined as $T_C \rightarrow (X(R)Y)$, where $X$ and $Y$ are temporal variables, and $R$ represents temporal relations between $X$ and $Y$. A temporal concept $c_t \in N_{c_t}$ can be formally expressed as $c_t = \diamond(X)\overline{T}_C.C$, where $C$ denotes concept expressions consisting of atomic concepts of the form $A$ and features of the form $f$, $X$ is a temporal variable introduced by the $\diamond$ temporal existential quantifier, and $\overline{T}$ denotes a temporal constraint.

In $TL\text{-}\mathcal{SI}$, concept $C$ at interval $X$ is denoted as $C@X$. Actions can be described using temporal concepts by limiting concept expressions with temporal variables to an individual set at a particular interval.

The Tarski-style semantics of *TL-SI* is defined using a temporal structure (see Definition 4.40).

**Definition 4.40 (*TL-SI* Temporal Structure)** A linear, unbounded, and dense temporal structure is defined as $\tau = (\rho, <)$, where $\rho$ is a set of time points and $<$ is a strict partial order on $\rho$.

The *TL-SI* interpretation is defined in Definition 4.41.

**Definition 4.41 (*TL-SI* Interpretation)** A *TL-SI* interpretation is defined as $\mathcal{I} = \left(\tau_<^*, \Delta^{\mathcal{I}}, \cdot^{\mathcal{I}}\right)$, where $\tau_<^*$ is the interval set of temporal structure $\tau$, $\Delta^{\mathcal{I}}$ is the object domain, and $\cdot^{\mathcal{I}}$ is a primitive interpretation function, which maps every concept $C \in N_c \cup N_{c_t}$ to the subset of $\tau_<^* \times \Delta^{\mathcal{I}}$, every atomic feature $f^{\mathcal{I}} : \left(\tau_<^* \times \Delta^{\mathcal{I}}\right)$ and atomic parametric feature $*g^{\mathcal{I}} : \Delta^{\mathcal{I}}$ to a subset of $\Delta^{\mathcal{I}}$, and every role $R \in N_R$ to a subset of $\Delta^{\mathcal{I}} \times \Delta^{\mathcal{I}}$.

#### 4.2.7.4   *DL-Lite*-Based Temporal Description Logics

Members of the *DL-Lite* family have been extended with temporal constructors with varying levels of expressivity. Beyond the usual $\mathsf{U}$ (*until*) and $\mathsf{S}$ (*since*) operators, these description logics utilize the $\circ_F$ (*next time*) and $\circ_P$ (*previous time*) operators used in quantitative evolution constraints, as well as the $\square_F$ (*always in the future*) and $\square_P$ (*always in the past*) operators required for qualitative evolution constraints, which are defined as follows: $\circ_F = \bot \mathsf{U} C$, $\circ_P = \bot \mathsf{S} C$, $\square_F C = \neg \circ_F \neg C$, $\square_P C = \neg \circ_P \neg C$. The temporal operator $\boxplus$ (read as *always* or *at all—past, present, and future—time instants*) represents timestamping constraints and is defined as shown in Eq. (4.2).

$$(\boxplus C)^{\mathcal{I}(n)} = \bigcap_{k \in Z} C^{\mathcal{I}(k)} \tag{4.2}$$

Using these temporal operators, the *DL-Lite*-based temporal description logic names reflect their temporal concept restrictions as follows:

- $\mathrm{T}_U$ allows temporal concepts of the form $\mathbf{D} ::= C \mid \boxplus C$
- $\mathrm{T}_{FP}$ restricts temporal concepts to the form $\mathbf{D} ::= C \mid \square_F C \mid \square_P C$
- $\mathrm{T}_{FPX}$ supports temporal concepts of the form $\mathbf{D} ::= C \mid \square_F C \mid \square_P C \mid \circ_F C \mid \circ_P C$

By allowing different sets of the above temporal constructors and concept restrictions, 12 temporal description logics can be differentiated (see Table 4.15) [141].

These variations are intended for different applications. For example, $\mathrm{T}_{FP}DL\text{-}LiteN_{bool}$ cannot express quantitative evolution constraints, just timestamping and qualitative evolution constraints, while $\mathrm{T}_{FPX}Lite_{bool}^N$ can express all three.

**Table 4.15** *DL-Lite*-based temporal description logics

| | | Temporal constructors | | |
| --- | --- | --- | --- | --- |
| | | $U/S$, $\bigcirc_F/\bigcirc_P$, $\square_F/\square_P$ | $\square_F/\square_P$ | $\boxplus$ |
| Concept inclusions | Bool | $T_{US}DL\text{-}Lite_{bool}^N$ <br> $T_{FPX}DL\text{-}Lite_{bool}^N$ | $T_{FP}DL\text{-}Lite_{bool}^N$ | $T_U DL\text{-}Lite_{bool}^N$ |
| | Krom | $T_{FPX}DL\text{-}Lite_{krom}^N$ | $T_{FP}DL\text{-}Lite_{krom}^N$ | $T_U DL\text{-}Lite_{krom}^N$ |
| | Core | $T_{FPX}DL\text{-}Lite_{core}^N$ | $T_{FP}DL\text{-}Lite_{core}^N$ | $T_U DL\text{-}Lite_{core}^N$ |
| Temporalized roles | | $T_X^R DL\text{-}Lite_{bool}^N$ | – | $TR_U DL\text{-}Lite_{bool}^N$ |

#### 4.2.7.5   $TDL_{BR}^{\star}$

The $TDL_{BR}^{\star}$ is a tense-extensible description logic aligned with *XBRL* (eXtensible Business Reporting Language) [142]. $TDL_{BR}^{\star}$ concept declarations utilize atomic concept, negation, and intersection definitions as usual, but $TDL_{BR}^{\star}$ concepts can also be atomic and complex temporal concepts, concepts in a particular tense, and multiconstraints (see Definition 4.42).

**Definition 4.42 ($TDL_{BR}^{\star}$ Concept)** In $TDL_{BR}^{\star}$, concepts are defined as $C ::= \top_1$ $|A|\neg C|C_1 \sqcap C_2|AT|T|C@t|(\leqslant k[i]R)$, where $\top_1$ denotes full concept, $A$ denotes atomic concept, $AT$ denotes atomic temporal concept, $T$ denotes temporal concept, $C@t$ denotes the concept in tense $t$, and $\leqslant k[i]R$ is a multiconstraint related to the $i$th element of relationship $R$, where $i$ is either 1 or 2 and $k$ is a nonnegative integer.

Beyond the standard object domain and interpretation function, a $TDL_{BR}^{\star}$ interpretation also incorporates a temporal interpretation set and a temporal interpretation function (see Definition 4.43).

**Definition 4.43 ($TDL_{BR}^{\star}$ Interpretation)** A $TDL_{BR}^{\star}$ interpretation is defined as $\mathcal{I} = (\Delta^{\mathcal{I}}, \Delta T, \cdot^{\mathcal{I}}, \cdot^{T})$, where $\Delta^{\mathcal{I}}$ denotes a nonempty set, $\Delta T$ denotes a temporal interpretation set, $\cdot^{\mathcal{I}}$ is the interpretation function, and $\cdot^{T}$ is the temporal interpretation function.

#### 4.2.7.6   DL-CTL

DL-CTL extends description logics with propositional branch-time logic (CTL) and was designed to specify temporal properties on state transition systems [143]. The concepts and concept expressions are defined similarly in DL-CTL than in conventional description logics; however, DL-CTL formulae are constructed with the temporal operators of CTL (see Definition 4.44).

**Definition 4.44 (DL-CTL formula)** A DL-CTL formula is defined as $\phi, \psi ::=$ $C \sqsubseteq D \mid C(p) \mid R(p, q) \mid \neg\phi \mid \phi \wedge \psi \mid EX\phi \mid AF\phi \mid E(\phi U\psi)$, where $p, q \in N_I, R \in N_R,$ $C, D \in N_C$.

DL-CTL has similar semantics to CTL (based on the structure in which the states are organized as a branch structure—see Definition 4.45), but unlike CTL, the state in DL-CTL is mapped to DL interpretations.

**Definition 4.45 (DL-CTL structure)** A DL-CTL structure is a quadruple of the form $M = (S, T, \Delta, \mathcal{I})$, where $S$ is a set of all states, $T \subseteq S \times S$ is a binary relation (transition) between states, $\Delta$ is the interpretation domain, and for every state $s \in S$, function $\mathcal{I}$ provides $s$ with a DL interpretation of the form $\mathcal{I}(s) = (\Delta, \cdot^{\mathcal{I}(s)})$, such that for every concept $C_i \in N_C$, there is $C_i^{\mathcal{I}(s)} \subseteq \Delta$; for every role name $R_i \in N_R$, $\exists R_i^{\mathcal{I}(s)} \subseteq \Delta \times \Delta$; for every individual name $I_i \in N_I$, $\exists I_i^{\mathcal{I}(s)} \in \Delta$; and for every state $s' \in S$, $\exists p_i^{\mathcal{I}(s)} = p_i^{\mathcal{I}(s')}$.

#### 4.2.7.7  The Temporal Extension of $\mathcal{SHIN}^{(\mathcal{D})}$

The temporal extension of $\mathcal{SHIN}^{(\mathcal{D})}$ which, when implemented in OWL, is called *tOWL*, provides time points, relations between time points, intervals, and timeslices [144]. It is particularly useful for representing complex temporal aspects, such as process state transitions [145].

#### 4.2.7.8  HS-Lite$_{horn}^{H}$

The combination of the Halpern-Shoham interval temporal logic (*HS* logic) with the *DL-Lite$_{horn}^{H}$* description logic is called *HS-Lite$_{horn}^{H}$* [146]. The temporal operators of the *HS* logic are of the form $\langle R \rangle$ (corresponds to $\diamond$) or $[R]$ (corresponds to $\Box$), where R is one of Allen's interval relations. The propositional variables of *HS* are interpreted by sets of closed intervals of the form $[i,j]$ of some flow of time (e.g., $\mathbb{Z}$, $\mathbb{R}$).

*HS-Lite$_{horn}^{H}$* represents temporal data through assertions, while temporal concept and role inclusions are used to impose constraints on the data and introduce new concepts and roles. The language of the *HS-Lite$_{horn}^{H}$* description logic contains atomic concepts, atomic roles, and individual names. Basic concepts $B$, basic roles $R$, temporal concepts $C$, and temporal roles $S$ are given by the following grammar: $R ::= P_k | P_k^-, B ::= A_k | \exists R, S ::= R | [R]S | \langle R \rangle S$, and $C ::= B | [R]C | \langle R \rangle C$, where R is one of Allen's interval relations or the universal relation $G$.

An *HS-Lite$_{horn}^{H}$* TBox is a finite set of concept and role inclusions and disjointness constraints of the form $C_1 \sqcap \ldots \sqcap C_k \sqsubseteq C^+, C_1 \sqcap \ldots \sqcap C_k \sqsubseteq \bot, S_1 \sqcap \ldots \sqcap S_k \sqsubseteq S^+, S_1 \sqcap \ldots \sqcap S_k \sqsubseteq \bot$, where $C^+$ and $R^+$ denote temporal concepts and roles without the $\diamond$ operator. An *HS-Lite$_{horn}^{H}$* ABox is a finite set of atoms of the form $A_k(a,i,j)$ or $P_k(a,b, i,j)$, where temporal constants $i \leqslant j$ are given in binary.

An *HS-Lite$_{horn}^{H}$* interpretation $\mathcal{I}$ consists of a family of standard (atemporal) description logic interpretations $\mathcal{I}[i,j] = (\Delta^{\mathcal{I}}, \cdot^{\mathcal{I}[i,j]})$, for all $i, j \in \mathbb{Z}$ with $i \leqslant j$, where $\Delta^{\mathcal{I}} \neq \emptyset$, $a_k^{\mathcal{I}[i,j]} = a_k^{\mathcal{I}}$ for some (fixed) $a_k^{\mathcal{I}} \in \Delta^{\mathcal{I}}$, $A_k^{\mathcal{I}[i,j]} \subseteq \Delta^{\mathcal{I}}$ and $P_k^{\mathcal{I}[i,j]} \subseteq \Delta^{\mathcal{I}} \times \Delta^{\mathcal{I}}$. The role and concept constructors are interpreted in $\mathcal{I}$ as shown in Eqs. (4.3)–(4.6).

$$(P_k^-)^{\mathcal{I}[i,j]} = \left\{ (x,y) | (y,x) \in P_k^{\mathcal{I}[i,j]} \right\} \tag{4.3}$$

$$(\exists R)^{\mathcal{I}[i,j]} = \left\{x | (x,y) \in R^{\mathcal{I}[i,j]}\right\} \qquad (4.4)$$

$$([R]S)^{\mathcal{I}[i,j]} = \bigcap_{[i,j]R[i',j']} S^{\mathcal{I}[i',j']} \qquad (4.5)$$

$$([R]C)^{\mathcal{I}[i,j]} = \bigcap_{[i,j]R[i',j']} C^{\mathcal{I}[i',j']} \qquad (4.6)$$

The constructors are interpreted analogously for the $[i',j']$ diamond operator.

#### 4.2.7.9  IMPNL and MITDL

The *IMPNL* interval-based temporal description logic, inspired by *Metric Propositional Neighborhood Logic (MPNL)*, consists of a set *AP* of propositional variables, the $\neg$ (atomic negation), $\vee$ (or), and $\wedge$ (and) logical operators, and two temporal operators, $\diamond_r$, $\diamond_l$, which correspond to Allen's *meet* and *met-by* relations [147].

The combination of a restricted version of IMPNL (in which negation is not available) and the $\mathcal{ALC}$ description logic form the *Metric Interval-Based Temporal Description Logic (MITDL)*. MITDL is expressive enough to model not only static aspects (e.g., relations between concepts in a domain), but also dynamic aspects (e.g., time constraints).

An MITDL signature is a quintuple of the form $\Sigma = (\mathbf{C}, \mathbf{R}, \mathbf{I}, \top, \bot)$, where $\mathbf{C}$ is a set of concept names, $\mathbf{R}$ is a set of role names, $\mathbf{I}$ is a set of individual names, while $\top$ is the top and $\bot$ is bottom concept as usual. An MITDL formula $\psi$ over signature $\Sigma$ is defined recursively as follows: $\psi = (C = \top)_k | C(a)_k | R(a,b)_k | \top_k | \bot_k | \psi_1 \vee \psi_2 | \psi_1 \wedge \psi_2 | \diamond_r \psi | \diamond_l \psi$, where $C$ represents atomic concepts, $R$ represents atomic roles, $a,b$ are individual names, and $k \in \mathbb{N}$.

The MITDL semantics are based on the structure $M = (\mathbb{D}, \mathbb{I}(\mathbb{D}), \mathbb{S})$, where the pair $(\mathbb{D}, \mathbb{I}(\mathbb{D}))$ is a strict interval structure, and $\mathbb{S}$ is a set of $\mathcal{ALC}$ interpretations over the intervals defined.

### 4.2.8  Spatiotemporal Description Logics

Despite the number of spatial and temporal description logics, there are very few mentions in the literature about spatiotemporal description logics, one of which is $\mathcal{ALCRP}(S_2 \oplus T)$. It extends $\mathcal{ALCRP}^{(\mathcal{D})}$ to provide formal declarations of concrete domains with not only space, but also with time [148]. In $\mathcal{ALCRP}(S_2 \oplus T)$, an appropriate concrete domain $S_2$ is defined for polygons using RCC8 relations as basic predicates of the concrete domain, while the temporal aspect is provided by the concrete domain $T$ defined as a set of time intervals and the 13 Allen interval relations. $S_2 \oplus T$ is the combination of $S_2$ and $T$, which can be used to define concrete spatiotemporal domains.

### *4.2.9  Fuzzy Description Logics*

Intelligent systems without well-defined boundaries or precisely defined criteria have to handle vagueness. Fuzzy description logics do just that by adding fuzzy logic to description logics. They can be implemented to ontologies to represent the fuzzy relationships between the context and depicted concepts of image and video contents, and perform both abductive and deductive reasoning over the axioms and rules [149].

The names of most fuzzy description logics start with f- written in normal text font (rather than the CMSY10 font), for example, f-$\mathcal{ALC}$, f-$\mathcal{ALCIQ}$, f-$\mathcal{SHIN}$, f-$\mathcal{SHOIN}$, f-$\mathcal{SHOIQ}$, and f-$\mathcal{SROIQ}$. Exceptions include $\mathcal{ALCQ}_F^+$, which includes fuzzy quantifiers, and C-$\mathcal{SHOIN}$, an extension of $\mathcal{SHOIN}$ based on a cloud model [150]. Some of the most common fuzzy description logics are described in the following sections.

#### 4.2.9.1  The f-$\mathcal{SHIN}$ Fuzzy Description Logic

The fuzzy version of the $\mathcal{SHIN}$ description logic is called f-$\mathcal{SHIN}$, which can be briefly defined as follows [151]. Consider an alphabet of distinct concept names (**C**), role names (**R**), and individual names (**I**). Let $N_R \in \mathbf{R}$ be an atomic role, $R$ a role, and $C, D$ concepts. The set of f-$\mathcal{SHIN}$ concepts is the smallest set such that every concept name $C \in N_C$ is an f-$\mathcal{SHIN}$ concept. Valid f-$\mathcal{SHIN}$ roles are defined by the abstract syntax $R :: = N_R \mid R^-$. If $C$ and $D$ are f-$\mathcal{SHIN}$ concepts, $R$ is an f-$\mathcal{SHIN}$ role, $S$ a simple f-$\mathcal{SHIN}$ role, and $p \in \mathbb{N}$, then $(C \sqcup D)$, $(C \sqcap D)$, $(\neg C)$, $(\forall R.C)$, $(\exists R.C)$, $(\geqslant pS)$, and $(\leqslant pS)$ are also f-$\mathcal{SHIN}$ concepts. The syntax and semantics of f-$\mathcal{SHIN}$ concepts are summarized in Table 4.16.

**Table 4.16**  Syntax and semantics of f-$\mathcal{SHIN}$ concepts

| Constructor | Syntax | Semantics |
|---|---|---|
| Top concept | $\top$ | $\top^{\mathcal{I}}(a) = 1$ |
| Bottom concept | $\bot$ | $\bot^{\mathcal{I}}(a) = 0$ |
| Negation | $\neg C$ | $(\neg C)^{\mathcal{I}}(a) = 1 - C^{\mathcal{I}}(a)$ |
| Disjunction | $C \sqcup D$ | $(C \sqcup D)^{\mathcal{I}}(a) = \max(C^{\mathcal{I}}(a), D^{\mathcal{I}}(a))$ |
| Conjunction | $C \sqcap D$ | $(C \sqcap D)^{\mathcal{I}}(a) = \min(C^{\mathcal{I}}(a), D^{\mathcal{I}}(a))$ |
| Universal quantification | $\forall R.C$ | $(\forall R.C)^{\mathcal{I}}(a) = \inf_{b \in \Delta^{\mathcal{I}}} \{ \max(1 - R^{\mathcal{I}}(a,b), C^{\mathcal{I}}(b)) \}$ |
| Existential quantification | $\exists R.C$ | $(\exists R.C)^{\mathcal{I}}(a) = \sup_{b \in \Delta^{\mathcal{I}}} \{ \min(R^{\mathcal{I}}(a,b), C^{\mathcal{I}}(b)) \}$ |
| At-least restriction | $\geqslant pR$ | $(\geqslant pR)^{\mathcal{I}}(a) = \sup_{b_1, \ldots, b_p \in \Delta^{\mathcal{I}}} \min_{i=1}^{p} (R^{\mathcal{I}}(a,b_i))$ |
| At-most restriction | $\leqslant pR$ | $(\leqslant pR)^{\mathcal{I}}(a) = \inf_{b_1, \ldots, b_p \in \Delta^{\mathcal{I}}} \max_{i=1}^{p+1} \{ 1 - R^{\mathcal{I}}(a,b_i) \}$ |
| Inverse role | $R^-$ | $(R^-)^{\mathcal{I}}(b,a) = R^{\mathcal{I}}(a,b)$ |

An f-$\mathcal{SHIN}$ TBox is a finite set of fuzzy concept axioms. An f-$\mathcal{SHIN}$ ABox is a finite set of fuzzy assertions of the form $\langle C(a) \bowtie n \rangle$, $\langle R(a,b) \bowtie n \rangle$, where $\bowtie$ stands for $\geqslant, >, \leqslant, <$, or $a \neq b$ for $a,b \in \mathbf{I}$. An f-$\mathcal{SHIN}$ RBox is a finite set of fuzzy role axioms. An f-$\mathcal{SHIN}$ interpretation is a pair, $\mathcal{I} = \langle \Delta^{\mathcal{I}}, \cdot^{\mathcal{I}} \rangle$, where the nonempty set $\Delta^{\mathcal{I}}$ is the object domain, and $\cdot^{\mathcal{I}}$ is a fuzzy interpretation function, which maps individual names of the form $a$ to elements of $a^{\mathcal{I}} \in \Delta^{\mathcal{I}}$, concept names of the form $A$ to a membership function $A^{\mathcal{I}} : \Delta^{\mathcal{I}} \to [0,1]$, and role names of the form $R$ to a membership function $R^{\mathcal{I}} : \Delta^{\mathcal{I}} \times \Delta^{\mathcal{I}} \to [0,1]$.

### 4.2.9.2  The f-$\mathcal{SHOIN}$ Fuzzy Description Logic

The fuzzy extension of the $\mathcal{SHOIN}$ description logic is called f-$\mathcal{SHOIN}$, which can be briefly defined as follows [152]. Consider an alphabet of concept names (**C**), abstract role names ($R_A$), concrete role names ($R_D$), abstract individuals ($I_A$), concrete individuals ($I_D$), and concrete datatypes (**D**). The set of $\mathcal{SHOIN}^{(\mathcal{D})}$ roles is defined by $R_A \cup \{R^- \mid R \in R_A\} \cup R_D$, where $R^-$ is called the inverse role of $R$. Let $A \in C, R, S \in R_A$, where $S$ is a simple role, $T_i \in R_D$, $d$ is a datatype, $o, o_1, \ldots, o_k \in I_A$, $c$, $n \in (0,1]$, $p \in \mathbb{N}$, and $k \in \mathbb{N}^*$. Based on these sets, the f-$\mathcal{SHOIN}^{(\mathcal{D})}$ concepts are defined inductively by the production rule shown in Definition 4.46.

**Definition 4.46 (f-$\mathcal{SHOIN}^{(\mathcal{D})}$ Concept)**  The f-$\mathcal{SHOIN}^{(\mathcal{D})}$ concepts are defined by the rule $C, D \to \top \mid \bot \mid A \mid C \sqcap D \mid C \sqcup D \mid \neg C \mid \exists R.C \mid \forall R.C \mid \geqslant pS \mid \leqslant pS \mid \exists T.u \mid \forall T.u \mid \geqslant pT \mid \leqslant pT \mid \{o\} \mid \{o_1, \ldots, o_k\} \mid R(o)u \to d \mid \{c\}$.

The semantics of f-$\mathcal{SHOIN}$ are provided by the fuzzy interpretation of the form $\mathcal{I} = (\Delta^{\mathcal{I}}, \cdot^{\mathcal{I}})$ and the interpretation of the concrete (datatype) domain $D = (\Delta_{\mathcal{D}}, \cdot^{\mathcal{D}})$ (see Definition 4.47).

**Definition 4.47 (f-$\mathcal{SHOIN}^{(\mathcal{D})}$ Interpretation)**  An f-$\mathcal{SHOIN}$ interpretation is defined by a quadruple of the form $\mathcal{I} = (\Delta^{\mathcal{I}}, \Delta_{\mathcal{D}}, \cdot^{\mathcal{I}}, \cdot^{\mathcal{D}})$, where the abstract domain $\Delta^{\mathcal{I}}$ is a nonempty set of objects, the datatype domain $\Delta_{\mathcal{D}}$ is the domain of interpretation of all datatypes (disjoint from $\Delta^{\mathcal{I}}$) consisting of data values, and $\cdot^{\mathcal{I}}$ and $\cdot^{\mathcal{D}}$ are two fuzzy interpretation functions, which map an abstract individual $a$ to an element $a^{\mathcal{I}} \in \Delta^{\mathcal{I}}$, a concrete individual $c$ to an element $c^{\mathcal{D}} \in \Delta_{\mathcal{D}}$, a concept name $A$ to a function $A^{\mathcal{I}} : \Delta^{\mathcal{I}} \to [0,1]$, an abstract role name $R$ to a function $R^{\mathcal{I}} : \Delta^{\mathcal{I}} \times \Delta^{\mathcal{I}} \to [0,1]$, a datatype $d$ to a function $d^{\mathcal{D}} : \Delta_{\mathcal{D}} \to [0,1]$, and a concrete role name $T$ to a function $T^{\mathcal{I}} : \Delta^{\mathcal{I}} \times \Delta_{\mathcal{D}} \to [0,1]$.

This makes it possible to assign any degree between 0 and 1 to any object or object pair of a fuzzy concept. Using this interpretation, the semantics of f-$\mathcal{SHOIN}$ concepts can be summarized as shown in Table 4.17.

**Table 4.17** Syntax and semantics of f-$\mathcal{SHOIN}^{(\mathcal{D})}$ concepts

| Constructor | Syntax | Semantics |
|---|---|---|
| Top concept | $\top$ | $\top^{\mathcal{I}}(a) = 1$ |
| Bottom concept | $\bot$ | $\bot^{\mathcal{I}}(a) = 0$ |
| Data value | $c$ | $c^{\mathcal{I}} = c^{\mathcal{D}}$ |
| Datatype | $d$ | $d^{\mathcal{I}}(y) = d^{\mathcal{D}}(y)$ |
| Conjunction | $C \sqcap D$ | $(C \sqcap D)^{\mathcal{I}}(a) = t(C^{\mathcal{I}}(a), D^{\mathcal{I}}(a))$ |
| Disjunction | $C \sqcup D$ | $(C \sqcup D)^{\mathcal{I}}(a) = u(C^{\mathcal{I}}(a), D^{\mathcal{I}}(a))$ |
| Negation | $\neg C$ | $(\neg C)^{\mathcal{I}}(a) = c(C^{\mathcal{I}}(a))$ |
| Nominal | $\{o\}$ | $\{o\}^{\mathcal{I}}(a) = 1$ iff $o^{\mathcal{I}} = a$, otherwise $\{o\}^{\mathcal{I}}(a) = o$ |
| One of | $\{o_1, \ldots, o_k\}$ | $\{o_1, \ldots, o_k\}^{\mathcal{I}}(a) = 1$ if $a \in \{o_1^{\mathcal{I}}, \ldots, o_k^{\mathcal{I}}\}$, 0 otherwise |
| Fills | $\exists R.\{o\}$ | $(\exists R.\{o\})^{\mathcal{I}}(a) = R^{\mathcal{I}}(a, o^{\mathcal{I}})$ |
| Existential quantification | $\exists R.C$ | $(\exists R.C)^{\mathcal{I}}(a) = \sup_{b \in \Delta^{\mathcal{I}}} t\left( R^{\mathcal{I}}(a, b), C^{\mathcal{I}}(b) \right)$ |
| Universal quantification | $\forall R.C$ | $(\forall R.C)^{\mathcal{I}}(a) = \inf_{b \in \Delta^{\mathcal{I}}} J\left( R^{\mathcal{I}}(a, b), C^{\mathcal{I}}(b) \right)$ |
| At-least restriction | $\geqslant pS$ | $(\geqslant pS)^{\mathcal{I}}(a) = \sup_{b_1, \ldots, b_p \in \Delta^{\mathcal{I}}} t\left( \overset{p}{\underset{i=1}{t}}, S^{\mathcal{I}}, (a, b_i), \underset{i<j}{t} \{b_i \neq b_j\} \right)$ |
| At-most restriction | $\leqslant pS$ | $(\leqslant pS)^{\mathcal{I}}(a) = \inf_{b_1, \ldots, b_{p+1} \in \Delta^{\mathcal{I}}} J\left( \overset{p+1}{\underset{i=1}{t}} S^{\mathcal{I}}(a, b_i), \underset{i<j}{u} \{b_i = b_j\} \right)$ |
| Inverse role | $R^-$ | $(R^-)^{\mathcal{I}}(b, a) = R^{\mathcal{I}}(a, b)$ |
| Datatype exists | $\exists T.d$ | $(\exists T.d)^{\mathcal{I}}(a) = \sup_{y \in \Delta_{\mathcal{D}}} t\left( T^{\mathcal{I}}(a, y), d^{\mathcal{I}}(y) \right)$ |
| Datatype value | $\forall T.d$ | $(\forall T.d)^{\mathcal{I}}(a) = \inf_{y \in \Delta_{\mathcal{D}}} J\left( T^{\mathcal{I}}(a, y), d^{\mathcal{I}}(y) \right)$ |
| Datatype at-least | $\geqslant pT$ | $(\geqslant pT)^{\mathcal{I}}(a) = \sup_{y_1, \ldots, y_p \in \Delta_{\mathcal{D}}} t\left( \overset{p}{\underset{i=1}{t}} R^{\mathcal{I}}(a, y_i), \underset{i<j}{t} \{y_i \neq y_j\} \right)$ |
| Datatype at-most | $\leqslant pT$ | $(\leqslant pT)^{\mathcal{I}}(a) = \inf_{y_1, \ldots, y_{p+1} \in \Delta_{\mathcal{D}}} J\left( \overset{p+1}{\underset{i=1}{t}} R^{\mathcal{I}}(a, y_i), \underset{i<j}{u} \{y_i \neq y_j\} \right)$ |
| Datatype nominal | $\{c\}$ | $\{c\}^{\mathcal{I}}(y) = 1$ iff $c^{\mathcal{D}} = y$, $\{c\}D(y) = o$ otherwise |

## 4.3 Extending DL Expressivity

Because most description logics can be considered as fragments of function-free first-order logic, it is straightforward to combine them with first-order Horn logic rules. This rule extension applied to OWL yields to the Semantic Web Rule Language (SWRL) discussed earlier in Chap. 2, which improves the expressivity of OWL. Reasoning over the combination of OWL and SWRL might be undecidable, however, which was the reason behind the introduction of expressive rule formalisms that are more restricted than SWRL. Improving the expressivity of description logics without sacrificing decidability has been researched since the introduction of OWL, resulting in complex formalisms such as the following:

- *Description Logic Programs (DLP)*: one of the approaches to retain decidability when extending OWL with SWRL rules is to identify the Horn logic rules directly expressible in OWL DL ( $\mathcal{SHOIN}$ ), which yields to Description Logic Programs (DLP) [153]
- *Description Logic Rules (DLR)*: a rule-based formalism (decidable SWRL fragments) for augmenting description logic knowledge bases [154]
- $\mathcal{SROIQV}$: the extension of $\mathcal{SROIQ}$ with nominal schemas that allow arbitrary DL-safe rules (as expressive as SWRL or RIF) to be expressed in the native syntax of the ontology language

## 4.4   Formal Representation of Images

Description logic-based formalisms are suitable for formally describing arbitrary image contents and sophisticated information related to entire images or regions of interest (ROIs) in images. Assume a photo of Rio de Janeiro with two ROIs, the Sugarloaf Mountain and Christ the Redeemer, and the aim is to describe the sculpture in more detail, such as by declaring its geocoordinates, material, height, and creators. This image can be represented using description logics as follows:

Statue(CHRISTTHEREDEEMER)
latitude(CHRISTTHEREDEEMER, "-22.951944444444443")
longitude(CHRISTTHEREDEEMER, "-43.21055555555556")
artform(CHRISTTHEREDEEMER, Sculpture)
artMedium(CHRISTTHEREDEEMER, Soapstone)
height(CHRISTTHEREDEEMER, "38")
creator(CHRISTTHEREDEEMER, PAULLANDOWSKI)
creator(CHRISTTHEREDEEMER, HEITORDASILVACOSTA)
contributor(CHRISTTHEREDEEMER, ALBERTCAQUOT)
contributor(CHRISTTHEREDEEMER, GHEORGHELEONIDA)
depicts(IMAGE1, RIODEJANEIRO)
partOf(ROI1, IMAGE1)
partOf(ROI2, IMAGE1)
depicts(ROI1, SUGARLOAFMOUNTAIN)
depicts(ROI2, CHRISTTHEREDEEMER)

The regions of interest can be declared using Media Fragment URI 1.0 identifiers,[27] where the URL ending identifies the top left corner coordinates, the width, and the height of the region (see Fig. 4.6).

This example is a 22-megapixel photo taken with a Canon EOS 5D Mark III with a resolution of $5760 \times 3840$, so that every $x,y$ coordinate between 0,0 and 5759,3839 uniquely identifies a pixel in the image. In this case, the top left coordinate of region 1 is 0,1390 and the top left coordinate of region 2 is 3699,620.

---

[27]https://www.w3.org/TR/media-frags/

**Fig. 4.6** Regions of interest to be identified using Media Fragment URIs

The Turtle serialization of Listing 4.1 yields to an LOD-enriched semantic description of the image and its two regions of interest (see Listing 4.1).

**Listing 4.1** Structured image and ROI description with rich semantics

```
@prefix dbpedia: <http://dbpedia.org/resource/> .
@prefix foaf: <http://xmlns.org/foaf/0.1/> .
@prefix geo: <http://www.w3.org/2003/01/geo/wgs84_pos#> .

<http://example.com/rio.jpg> foaf:depicts
dbpedia:Rio_de_Janeiro .
<http://example.com/rio.jpg#xywh=0,1390,2326,1228>
foaf:depicts dbpedia:Sugarloaf_Mountain .
<http://example.com/rio.jpg#xywh=3699,620,1594,2253>
foaf:depicts dbpedia:Christ_the_Redeemer_(statue) .
dbpedia:Christ_the_Redeemer_(statue) a dbpedia:Statue ;
geo:lat "-22.951944444444443" ; geo:long "-43.21055555555556"
; schema:artform dbpedia:Sculpture ; schema:artMedium
dbpedia:Soapstone ; schema:height "38" ; schema:creator
dbpedia:Paul_Landowski , dbpedia:Heitor_da_Silva_Costa" ;
schema:contributor dbpedia:Albert_Caquot" ,
dbpedia:Gheorghe_Leonida ; owl:sameAs
wikidata:Christ_the_Redeemer_(statue) .
```

## 4.5    Formal Representation of 3D Models and Scenes

To demonstrate the formal representation of 3D models, assume a 3D model of an Aston Martin Vantage created in AutoDesk 3ds Max to be described by 3D object features, such as geometry, shape, diffuse color, specular color, transparency, and so on. Such a model can be represented using description logics as follows:

3DModel(ASTONMARTINMODEL)
depicts(ASTONMARTINMODEL, ASTONMARTINVANTAGE)
skyColor(ASTONMARTINMODEL, 0.000 0.000 0.000)
translation(ASTONMARTINMODEL, -120.555 350.908 -140.466)
diffuseColor(ASTONMARTINMODEL, 0.910 0.902 0.859)
specularColor(ASTONMARTINMODEL, 1.000 1.000 1.000)
transparency(ASTONMARTINMODEL, 0.000)
shininess(ASTONMARTINMODEL, 0.079)
coordIndex(ASTONMARTINMODEL, 0 1 2 -1 3 4 5 -1 6 7 8 -1 9 10 11 -1 12 13 14 -1 15 16 17 -1 18 19 20 -1 21 22 23 -1 24 25 26 -1 27 28 29 -1 30 31 32 -1 33 34 35 -1 36 37 38 -1 39 40 41 -1 42 43 44 -1 45 46 . . .)
. . .

Using an X3D plugin, the 3D model can be exported to X3D (see Fig. 4.7).

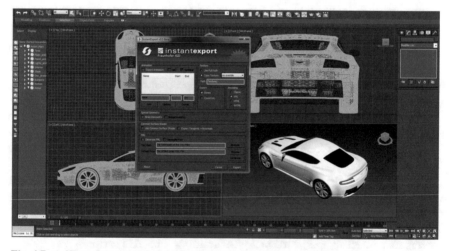

**Fig. 4.7**  A 3D model created in 3ds Max can be exported to semistructured X3D annotation

By serializing the description logic representation of the above model in Turtle and utilizing the 3D Modeling Ontology [155], a structured and semantically enriched representation of the model can be obtained (see Listing 4.2).

**Listing 4.2**  Fragment of the Turtle serialization of the 3D model in Fig. 4.7

```
@prefix foaf: <http://xmlns.com/foaf/0.1/> .
@prefix dbpedia: <http://dbpedia.org/resource/> .
@prefix 3d: <http://3dontology.org/> .

<http://3dontology.org/3dmodels/astonmartin/> a 3d:3DModel ,
foaf:depicts dbpedia:Aston_Martin_Vantage_(2005) ;
3d:createdIn 3d:AutoDesk3dsMax ;
3d:baseForm 3d:Polyhedron ; 3d:hasCompound 3d:Box ,
3d:Cylinder ; 3d:hasFaces "618210"^^xsd:nonNegativeInteger ;
3d:hasVertices "399929"^^xsd:nonNegativeInteger ;
3d:hasEdges "1854630"^^xsd:nonNegativeInteger ;
3d:skyColor "0.000 0.000 0.000" ;
3d:translation "-120.555 350.908 -140.466" ;
3d:diffuseColor "0.910 0.902 0.859" ;
3d:specularColor "1.000 1.000 1.000" ;
3d:transparency "0.000" ;
3d:shininess".079" ;
3d:coordIndex "0 1 2 -1 3 4 5 -1 6 7 8 -1 9 10 11 -1 12 13 14
-1 15 16 17 -1 18 19 20 -1 21 22 23 -1 24 25 26 -1 27 28 29
-1 30 31 32 -1 33 34 35 -1 36 37 38 -1 39 40 41 -1 42 43 44 -1
45 46 ..." .
```

Using the 3D Modeling Ontology, the geometric base form of the model is declared as a polyhedron, which is modeled as an editable poly in 3ds Max. In this example, two compounds are declared, but more can be added analogously. The number of faces and vertices also contributes to the formal description of the geometry of the model. Further properties include various color properties, transparency, and shininess as examples.

The result can be deployed in any other RDF serialization. This way, 3D models displayed on websites can be semantically enriched and described in HTML5 Microdata, JSON-LD, or RDFa, so that they can be indexed and retrieved by search engines not only based on textual description on the page, but also by characteristics of the 3D models themselves. Moreover, intelligent services can be provided via reasoning over the formally represented 3D models, as, for example, to compare two 3D models by volume or to find 3D models that contain an identical part.

## 4.6   Formal Representation of Audio

The semantic description of music files can provide structured metadata and high-level information about the content of a particular song, musical performance, or recording, and low-level descriptors to identify a song fragment or automatically locate similar items of audio databases. As described in Chap. 3, there are several vocabularies and ontologies to describe music, one of which is the ontology of MusicBrainz, one of the most comprehensive open content music databases. Low-level audio features, such as starting frequency (see Fig. 4.8), can be described using MPEG-7, and music metadata (title, composer, performer, etc.) using Dublin Core or specific music ontologies.

Using the ontology of MusicBrainz, song tracks can be annotated as `music-brainz:Track` (which is a subclass of `mpeg-7:AudioSegment`), and their music features defined with MPEG-7 terms, such as `mpeg-7:Beat`, which annotates the tempo of the music. All audio files can be uniquely identified by their *acoustic fingerprint* (audio fingerprint), a string deterministically generated from audio signals [156] using software such as Chromaprint,[28] and often visualized as an image (see Fig. 4.9).

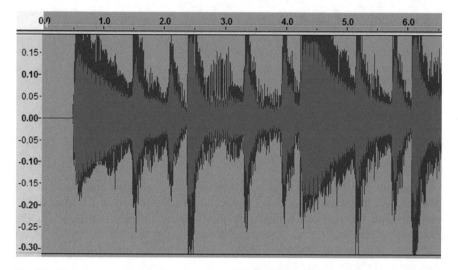

**Fig. 4.8** Waveform fragment of Elvis Presley's "Can't Help Falling in Love"

**Fig. 4.9** AcoustID of "Can't Help Falling in Love" by Elvis Presley (rotated counterclockwise by 90°)

Acoustic fingerprints allow files to be identified by the actual music content even without associated metadata; however, their standardization is yet to come.[29]

---

[28]https://acoustid.org/chromaprint

[29]MusicBrainz has used three audio fingerprinting systems since its release: PUID, TRM (TRM Recognizes Music), and AcoustID. Both the early RDFS and more recent OWL ontology file of MusicBrainz (neither of which is associated with the namespace URL, which is just a symbolic link) define TRM identifiers only (`musicbrainz:trmid`), which, however, are not supported by the MusicBrainz server since 2008. While currently AcoustID is provided for all the music tracks available on MusicBrainz, there is no property declared for AcoustID in the MusicBrainz ontology, and hence the fingerprint is omitted from the Turtle example here (Listing 4.3).

As an example, the description logic representation of Elvis Presley's "Can't Help Falling in Love" can be written as:

MusicRecording(TRACK1)
title(TRACK1, CANTHELPFALLINGINLOVE)
MusicArtist(ELVISPRESLEY)
memberOf(ELVISPRESLEY, THEJORDANAIRES)
byArtist(TRACK1, THEJORDANAIRES)
duration(TRACK1, 2M59S)
inAlbum(TRACK1, BLUEHAWAII)
dateCreated(TRACK1, 1961-03-23)
locationCreated(TRACK1, HOLLYWOOD)
genre(TRACK1, SOFTROCK)
inLanguage(TRACK1, AMERICANENGLISH)
key(TRACK1, DMAJOR)
tempo(TRACK1, 102)
startingFrequency(TRACK1, 3292.924805)
acoustid(TRACK1, 00c55f5b-5d97-460c-9410-b3979f34833a)

As usual, the above description logic formalism can be expressed in any RDF serialization. Listing 4.3 shows the Turtle serialization of the same song.

**Listing 4.3**  Music description in Turtle

```
@prefix dc: <http://purl.org/dc/elements/1.1/> .
@prefix dbpedia: <http://dbpedia.org/resource/> .
@prefix keys: <http://purl.org/NET/c4dm/keys.owl#> .
@prefix mpeg-7: <http://mpeg7.org/> .
@prefix mo: <http://purl.org/ontology/mo/> .
@prefix musicbrainz: <http://musicbrainz.org/ns/mmd-1.0#> .
@prefix schema: <http://schema.org/> .

dbpedia:Elvis_Presley a schema:Person , mo:MusicArtist .
<http://example.org/track1.mp3> a mpeg-7:Audio ,
schema:MusicRecording , musicbrainz:Track ;
dc:title "Can't Help Falling in Love" ;
schema:byArtist dbpedia:Elvis_Presley ,
dbpedia:The_Jordanaires ;
musicbrainz:duration "PT02M59S"^^xsd:duration ;
schema:inAlbum dbpedia:Blue_Hawaii_(Elvis_Presley_album) ;
schema:dateCreated "1961-03-23"^^xsd:date ;
schema:locationCreated dbpedia:Hollywood ;
schema:genre dbpedia:Soft_rock ;
schema:inLanguage "en-US" ;
mo:key keys:DMajor ;
mpeg-7:Beat "102"^^xsd:nonNegativeInteger ;
mpeg-7:StartingFrequency "3292.924805"^^xsd:float .
```

In this example, the duration and creation date are declared in the standard ISO 8601 format, and the language code in the standard IETF BCP 47 format.

While it is a best practice to reuse standard and de facto standard terms from existing ontologies rather than defining them, there are overlapping definitions, although they are not always identical and might provide a different level of specificity. For example, there are multiple options to declare a music recording, such as schema:MusicRecording, musicbrainz:Track, and mo:Track. A solo musician declared using mo:MusicArtist is more precise than the more generic schema:MusicGroup, which can be a solo musician or a music group, such as a band, an orchestra, or a choir, by definition.[30] Similarly, schema:byArtist is used in the above example because it is more specific than the widely used dc:author. Here mpeg-7:Beat is used to declare the tempo of the recording (102), which might seem similar to af:Beat (from the Audio Features Ontology) by name; however, the latter is a beat event defined as a point in time, not the tempo of the song.

## 4.7  Formal Representation of Video Scenes

The conceptualization of video contents with multimedia ontologies based on logical formalisms is suitable for video indexing, scene interpretation, and video understanding [157]. The formal description of video events is a major step towards bridging the semantic gap in content-based video retrieval, because it addresses some of the limitations of audio and video signal processing complemented by hard-coded concept mapping [158]. Most structured video representations are expressed in RDF to describe machine-readable statements in the form of subject-predicate-object triples, e.g., scene-depicts-person. The corresponding concepts are defined in controlled vocabularies written in RDFS or OWL, or OWL ontologies. Other structured data sources, such as knowledge bases and LOD datasets, provide machine-readable statements and factual data related to the depicted concepts. For example, the formal description of a video depicting a person relies on formalisms that support temporal segmentation to identify the corresponding video segment, spatial segmentation to identify the relevant regions of interest, and vocabularies or ontologies that provide the machine-readable definition of video clips and the depicted concepts, along with their properties. For example, Schema.org is suitable for declaring the director, file format, language, encoding, and other properties of video clips (schema:Clip). The "depicts" relationship is defined by the Friend of a Friend (FOAF) vocabulary (foaf:depicts). The definition of "Person" can be used from schema:Person or foaf:Person, both of which define typical properties of persons, including, but not limited to, name, gender, birthdate, and nationality. Frame-level concept annotation alone is not adequate to describe video scenes, which can be addressed by utilizing spatial and temporal description logics, and employing rule-based formalisms.

---

[30]https://schema.org/MusicGroup

### 4.7.1 Spatial Annotation

Ontologies that describe relevant low-level features, such as motion trajectory, are suitable for the spatial annotation of images and video frames (e.g., MPEG-7-based ontologies, such as Rhizomik).[31]

The geometry of 3D objects in CGI and computer animations can be described using 3D ontologies, such as the aforementioned X3D-based 3D Modeling Ontology. However, knowledge representation formalisms alone are not suitable for complex spatial annotation of unconstrained video scenes. The frame-level spatial annotation of videos can be achieved using Media Fragment URI 1.0 identifiers, which are suitable for annotating ROIs of video frames and video scenes, similar to images.

### 4.7.2 Temporal Annotation

General-purpose ontologies are *synchronic* (refer to a single point in time), and thus are unable to describe the temporal dimension of objects, such as changes and movements. In the multimedia domain, relationships between objects are *diachronic* (vary over time), which can be described with logical formalisms and ontologies, and temporal extensions of OWL.

One of the earliest implementations of description logics for shot-level video description utilized binary relations for transitions (type, to-shot), cameras (focus, size, motion,[32] to-segment), actions (name, agent, theme), actors (name, position, motion, gaze, appearance, audio), places (name, background, audio), and dialogs (speaker, onscreen, topic, transcript) [159]. Using this approach, some roles can be annotated automatically (using detection and classification algorithms), semiautomatically (e.g., dialogs, shot transitions), or manually (e.g., actors' names, action names, place names). Based on the above vocabulary, core video shot components can be described with description logics as follows:

Transition(TRANSITION1)
type(TRANSITION1, CUT)
to-shot(TRANSITION1, SHOT2)
Camera(CAMERA1)
focus(CAMERA1, SEA)
size(CAMERA1, long-shot)
motion(CAMERA1, pan-right)

---

[31]However, not all OWL mappings of MPEG-7 contain the necessary classes and relations.

[32]According to professional video camera operation terminology: dolly forward, dolly backward, pedestal up, pedestal down, boom up, boom down, track left, track right, pan left, pan right, tilt up, tilt down, zoom in, zoom out.

Segment(SEGMENT2)
to-segment(CAMERA1, SEGMENT2)
Place(LOCATION1)
name(LOCATION1, ADELAIDE)
background(LOCATION1, JETTY)
soundtrack(LOCATION1, SEAGULL)
Dialog(DIALOG1)
speaker(DIALOG1, BEN)
topic(DIALOG1, TRAVEL)

This can then be serialized in Turtle as shown in Listing 4.4.

**Listing 4.4** Temporal description in Turtle

```
@prefix : <http://example.com/> .
@prefix rdf: <http://www.w3.org/1999/02/22-rdf-syntax-ns#> .

:Transition1 a :Transition ; a :cut ; :to-shot :shot2 .
:Camera1 a :Camera ; :focus :sea ; :size :long-shot ; :motion
:pan-right ; :to-segment :Segment2 .
:Location1 a :Place ; :name :Adelaide ; :background :Jetty ;
:soundtrack :Seagull .
:Dialog1 a :Dialog ; :speaker :Ben ; :topic :Travel .
```

The *SWRL Temporal Ontology*[33] defines a valid-time temporal model to be used for modeling complex interval-based temporal information in OWL ontologies. This valid-time temporal model provides consistent representation for temporal information, where each fact (proposition) is associated with instants or intervals denoting the time or times when the fact holds. These temporal facts have a value and one or more valid times. In this model, every temporal fact is held to be true or valid during the time or times associated with the fact, and conclusions can be made only about facts for time periods within their valid time. All the entities of this model, including propositions, valid times (both instants and intervals), granularity, and duration, are defined in the form of OWL classes, object properties and datatype properties, restrictions, and nominals. A granularity is specified using the `Granularity` class, which has the following instances: `Years`, `Months`, `Days`, `Hours`, `Minutes`, `Seconds`, `Milliseconds`. Durations are defined by the `Duration` class. The `ValidTime` class defines the valid times. The `hasGranularity` property specifies the accuracy of the temporal information; the range of this property is `Granularity`. Instants and intervals can be declared using the two subclasses of the `ValidTime` class, namely, `ValidInstant` and `ValidPeriod`. These classes enable the representation of activities that occur at a single instant in time and activities that happen over an interval of time,

---

[33]http://swrl.stanford.edu/ontologies/built-ins/3.3/temporal.owl

respectively. If an activity occurs only once, the property will be functional. Activities that reoccur will use a nonfunctional property.

The valid times can be associated with the `ExtendedProposition` class representing a fact or proposition to model entities that can extend over time. The `hasValidTimes` property of the `ExtendedProposition` class can be used to save the valid times of the propositions. Instances of the `ExtendedProposition` class might have instants or intervals as values of the `hasValidTimes` property.

The SWRL Temporal Ontology also provides a set of SWRL built-ins than can be used to reason with temporal information defined using its temporal model (see Chap. 6).

The *OWL-Time* ontology[34] of the W3C is an OWL ontology that represents temporal concepts through topological relations between instants and intervals with the temporal operators defined by Allen. However, the OWL-Time ontology alone is not always sufficient to represent the changes of an object or a video event, and special domain ontologies are needed, such as the *4D-Fluent Ontology* [160].

Diachronic knowledge representation with OWL is challenging even if only those binary relations are considered that vary over time, which can be addressed by (simplified) *fluents*, i.e., relations that hold within a certain time interval and do not hold in others.[35] The simplest representation of binary fluents adds a time argument to binary predicates. However, this approach leads to a ternary predicate, which cannot be described in OWL without reification, or an alternate approach suggested by the W3C [161]. Another way to represent fluents is using a metalogical predicate, which maps the relationship to a time interval. While this approach uses binary predicates only, it is not straightforward to express in OWL, because this requires the formalization of relationships over relationships, and can lead to undecidability.

In the 4D-Fluent Ontology, the `TimeSlice` class represents temporal entity parts linked to the `TimeInterval` class, a class of the time domain. Each entity is associated with an instance of the `TimeSlice` by the `tsTimeSliceOf` object property, the latter of which is connected to an instance of the `TimeInterval` by the `tsTimeInterval` property.

Another 4D-fluent-based language beyond the aforementioned tOWL is *SOWL* [162]. SOWL is also an extension of OWL-Time, and enables the representation of both static and dynamic information.

The *Video Structure Ontology* of Perperis et al. is suitable for the annotation of movies decomposed to consecutive audiovisual segments [163]. Each segment (`VSO:TemporalSegment`) conveys auditory or visual information, or both. The `VSO:AudioVisualSegment` subclass is suitable for annotating the results of low-level video analysis, which then can be used for multimedia reasoning. The Video Structure Ontology can be used in combination with domain ontologies, such as the *Movie Violence Ontology (MVO)* and the *Movie Visual Ontology (MVisO)*, to

---

[34]https://www.w3.org/2006/time

[35]The original, non-simplified definition of fluent is a function that maps objects and situations to truth.

provide sophisticated semantics for visual events and visual objects of videos (e.g., MVO:Fighting, MVO:ShotsExplosions, MVO:Screaming, MVO:Dialogue, MVO:ActionScenes, MVisO:Weapon, MVisO:Face, MVisO:Building, MVisO:Vehicle, MVisO:Fire, MVisO:HighActivity).

### 4.7.3  Spatiotemporal Annotation

As an example, consider the chariot race of the movie *Ben-Hur* with Judah Ben-Hur and Messala, portrayed by Charlton Heston and Stephen Boyd, respectively (Metro-Goldwyn-Mayer, 1959). The aim is to identify the video scene with temporal data, annotate the ROIs depicting Ben-Hur and his chariot as moving regions and the ROI depicting Messala as a still region with spatiotemporal segmentation, and describe the movie scene, the two movie characters, and the actors who played in the corresponding roles. Using description logic, the knowledge representation of this video scene can be formalized as follows:

Movie(BENHUR)
adaptedFrom(BENHUR, BENHURATALEOFTHECHRIST)
Scene ⊑ VideoSegment
Scene(CHARIOTRACE)
sceneFrom(CHARIOTRACE, BENHUR)
hasStartTime(CHARIOTRACE, 02:40:28)
duration(CHARIOTRACE, 00:04:40)
hasFinishTime(CHARIOTRACE, 02:45:08)
depicts(CHARIOTRACE, Chariot_Racing)
MovieCharacter(JUDAHBENHUR)
portrayedBy(JUDAHBENHUR, CHARLTONHESTON)
MovieCharacter(MESSALA)
portrayedBy(MESSALA, STEPHENBOYD)
partOf(CHARIOTRACEROI1, CHARIOTRACE)
partOf(CHARIOTRACEROI2, CHARIOTRACE)
partOf(CHARIOTRACEROI3, CHARIOTRACE)
partOf(CHARIOTRACEROI2, CHARIOTRACEROI1)
MovingRegion(CHARIOTRACEROI1)
MovingRegion(CHARIOTRACEROI2)
StillRegion(CHARIOTRACEROI3)
depicts(CHARIOTRACEROI1, CHARIOT)
depicts(CHARIOTRACEROI2, JUDAHBENHUR)
depicts(CHARIOTRACEROI3, MESSALA)

The terms required to describe the above scene can be obtained from ontologies such as MPEG-7, VidOnt, the SWRL Temporal Ontology, DBpedia, FOAF, and Schema.org, whose namespaces have to be declared and the corresponding prefixes used in the triples of RDF serializations. In Turtle, for example, the chariot race scene of *Ben-Hur* can be described as shown in Listing 4.5.

**Listing 4.5** Spatiotemporal description of a video scene in Turtle

```
@prefix dbpedia: <http://dbpedia.org/resource/> .
@prefix mpeg-7: <http://mpeg7.org/> .
@prefix temporal: <http://swrl.stanford.edu/ontologies/built-
ins/3.3/temporal.owl> .
@prefix schema: <http://schema.org/> .
@prefix vidont: <http://vidont.org/> .
@prefix xsd: <http://www.w3.org/2001/XMLSchema#> .

dbpedia:Ben-Hur_(1959_film) vidont:basedOn dbpedia:Ben-
Hur:_A_Tale_of_the_Christ .
<http://example.com/benhur.mp4> a mpeg-7:Video ,
schema:Movie ; vidont:filmAdaptationOf dbpedia:Ben-
Hur_(1959_film) .
<http://example.com/benhur.mp4#t=2:42:19,2:50:57> a mpeg-
7:VideoSegmentTemporalDecomposition ; vidont:Scene ;
vidont:sceneFrom <http://example.com/benhur.mp4> ;
temporal:hasStartTime "02:42:19"^^xsd:time ;
temporal:duration "PT08M38S"^^xsd:duration ;
temporal:hasFinishTime "02:50:57"^^xsd:time ; vidont:depicts
dbpedia:Chariot_racing .
dbpedia:Judah_Ben-Hur a vidont:MovieCharacter ;
vidont:portrayedBy dbpedia:Charlton_Heston .
dbpedia:Messala_(Ben-Hur) a vidont:MovieCharacter ;
vidont:portrayedBy dbpedia:Stephen_Boyd .
<http://example.com/benhur.mp4#t=2:42:19,2:50:57&xywh=153,334
,340,166> a
mpeg-7:VideoSegmentSpatioTemporalDecomposition ,
mpeg-7:MovingRegion ; vidont:depicts dbpedia:chariot .
<http://example.com/benhur.mp4#t=2:42:19,2:50:57&xywh=172,368
,61,106> a mpeg-7:VideoSegmentSpatioTemporalDecomposition
, mpeg-7:MovingRegion ; vidont:depicts dbpedia:Judah_Ben-
Hur .
<http://example.com/benhur.mp4#t=2:42:19,2:50:57&xywh=1060,302
,80,28> a mpeg-7:VideoSegmentSpatioTemporalDecomposition
, mpeg-7:StillRegion ; vidont:depicts dbpedia:Messala_(Ben-
Hur) .
```

Note that the MPEG-7 concepts used above are defined in an OWL mapping of MPEG-7, which has a stable namespace. MPEG-7 implements concepts as elements and defines the various multimedia segment types as `xsd:complexType`, i.e., the MPEG-7 terms ending in "Type" indicate proprietary datatypes.

In this example, the spatiotemporal segmentation is declared using Media Fragment URI 1.0 identifiers. The positions of the selected shots are specified in Normal Play Time format according to RFC 2326, which is the default time scheme for Media Fragment URIs. The movie characters and the chariot are represented by the top left corner coordinates and the dimensions of the minimum bounding boxes, as shown in Fig. 4.10.

The RDF triples of this example can be visualized as an RDF graph, i.e., a directed, labeled graph in which the nodes represent the resources and values, and the arrows assign the predicates (see Fig. 4.11).

Since the RDF graphs that share the same resource identifiers naturally merge together [164], interlinking LOD concepts and individuals (e.g., `dbpedia:Judah_Ben-Hur`) makes the above graph part of the LOD Cloud.

**Fig. 4.10** Spatial annotation of ROIs with the top left corner coordinates and dimensions to be used by Media Fragment URIs. Movie scene by Metro-Goldwyn-Mayer

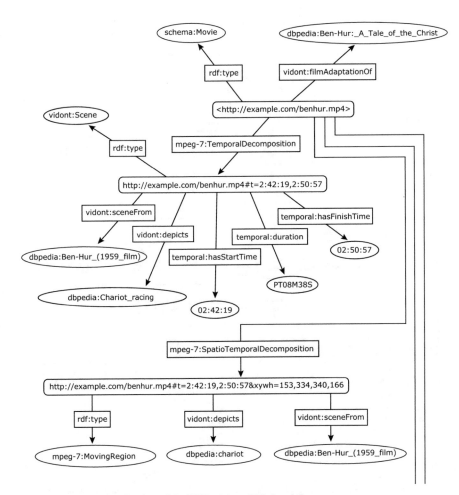

**Fig. 4.11** Graph visualization of the RDF triples of Listing 4.5

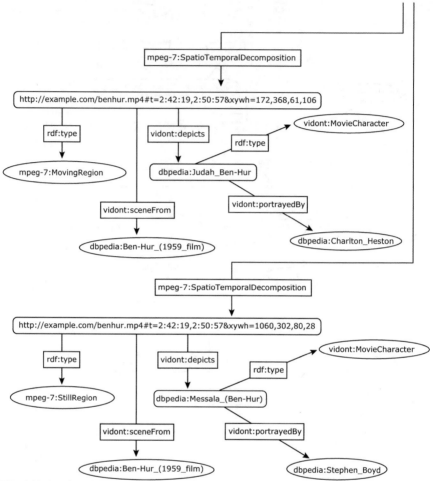

**Fig. 4.11** (continued)

## 4.8  Summary

This chapter introduced the formal grounding of web ontologies with description logics, a family of knowledge representation languages. Description logics enable the creation of ontologies with well-understood computational properties and favorable trade-offs between expressivity and scalability. The naming and annotation conventions of description logics have been listed. This chapter has shown how the meaning of description logic concepts and roles can be defined by their model-theoretic semantics based on interpretations. The reader has become familiar with the most common general-purpose description logics, as well as spatial and temporal description logic formalisms.

# Chapter 5
# Multimedia Ontology Engineering

The methods and methodologies for building ontologies, called ontology engineering, include ontology development, ontology transformation, ontology merging, ontology integration, ontology mapping, and ontology alignment. During ontology development, the knowledge domain and the scope to cover are determined, and the relevant concepts and their properties collected while considering the vocabulary of existing ontologies, followed by the creation of the concept and role hierarchy with formal grounding. For very expressive ontologies, rules are also defined to complement the DL axioms. The integrity of the resulting ontology should be checked with reasoners. This chapter lists both the best practices and the powerful software tools used by ontology engineers to perform the above tasks correctly and efficiently.

## 5.1 Introduction to Ontology Engineering

Semantic Web ontologies provide a common and shared knowledge ready to be transmitted between applications, thereby achieving interoperability across organizations and software agents of different areas or different views of the same area. Ontologies enable domain knowledge reuse, make domain assumptions explicit, and separate domain knowledge from operational knowledge [165]. The field of computer science that covers the methods and methodologies for building ontologies is called *ontology engineering*. The terminology of ontology engineering extends the core modeling terms of description logics and OWL. Classes (concepts) are used similarly; however, properties (roles) are sometimes called *slots*, and the role restrictions are called *facets*. In ontology engineering, individuals are also called *instances*.

The typical workflow of creating ontologies (*ontology design*) includes the following steps: determining the domain and scope of the ontology, assessing potential term reuse from existing ontologies, enumerating the terms of the

© Springer International Publishing AG 2017                                    121
L.F. Sikos, *Description Logics in Multimedia Reasoning*,
DOI 10.1007/978-3-319-54066-5_5

knowledge domain, building the class hierarchy, defining the properties, and adding individuals and (optionally) rules. However, ontology design is not a linear process but an iterative one: any of the steps can be repeated several times to refine the representation of the knowledge domain of the ontology being developed. There are many ways to model a domain; the selected method depends on the anticipated audience and applications.

Not all ontologies are developed from scratch. *Ontology transformation* is the development of a new ontology to deal with the new requirements of an existing ontology for a new purpose. *Ontology merging* is the process of creating a new single coherent ontology from two or more ontologies of the same knowledge domain. *Ontology integration* is the creation of a new ontology from two or more source ontologies from different knowledge domains. *Ontology mapping* formalizes semantic relationships between entities from different ontologies. *Ontology alignment* is the process of creating a consistent and coherent link between two or more ontologies, where statements of the first ontology confirm the statements of the second ontology or further ontologies.

## 5.1.1   Specification

An important step in ontology engineering is to specify the knowledge domain to cover, the anticipated audience, how detailed the ontology definitions should be, and for what purpose the ontology will be used.

### 5.1.1.1   Determining the Knowledge Domain and Ontology Scope

Ontology scope is determined by *ontology granularity*. *General-purpose ontologies* and *common sense ontologies* cover generic human knowledge, and are often used in combination with other types of ontologies. *Upper ontologies*, also known as upper-level ontologies, top-level ontologies, or foundational ontologies, are generic ontologies applicable to various domains. *Domain ontologies* describe the vocabulary of a specific knowledge domain with a specific viewpoint and factual data. *Application ontologies* (local ontologies) are specialized domain ontologies without consensus and knowledge sharing, i.e., they represent one particular model of a knowledge domain according to a single viewpoint or user. *Core reference ontologies* are standardized or de facto standard ontologies used by different user groups to integrate their different viewpoints about a knowledge domain by merging several domain ontologies.

Determining the knowledge domain and scope of an ontology can be defined by answering the following questions:

- What is the field of interest or area of concern the ontology will cover?
- What will be the potential application areas of the ontology?

- What types of questions should the formal knowledge representation of the ontology provide answers for?
- Who is the anticipated audience?
- Who will maintain the ontology?

## 5.1.2   Knowledge Acquisition

The information formally represented in ontologies can be retrieved from different types of information sources via knowledge elicitation techniques or via knowledge extraction from text, thesauri, relational databases, or UML diagrams. Information can be retrieved from the text corpus using natural language processing algorithms. Thesauri are particularly useful for designing multilingual ontologies. In contrast to plain text, the information acquired from relational databases include not only concepts, but also semantic relations and properties [166]. UML is widely adopted; however, it does not provide formal semantics, so it has to be complemented by constraints on concepts using a language such as OCL (Object Constraint Language) [167] or by advanced methodologies (e.g., [168]).

## 5.1.3   Conceptualization

The conceptualization of a knowledge domain formalizes implicit human knowledge about the constrained portion of the domain as a semantic structure. A partial specification of this structure is the ontology. The conceptual model of an ontology can be created in a modeling language such as the Unified Modeling Language (UML). The logical underpinning of an ontology formally expresses the conceptualization in a description logic, and the axioms of this conceptualization are usually implemented in OWL, as shown in Chaps. 2 and 4. Consequently, conceptualization is language-independent, but ontologies are language-dependent. What is common in conceptualization and actual ontology implementations is that concepts usually represent physical or logical objects in the form of nouns, and the properties (roles) are expressed as verbs (e.g., possession verbs) or predicates of a simple sentence. OWL ontologies almost never express all possible constraints [169], and the level of detailing in the conceptualization is restricted according to the intended application or usage scenario.

## 5.1.4   Assessment of Potential Term Reuse

Standard and well-established de facto standard definitions should be reused in ontology engineering as per Semantic Web best practices [170]. Before enumerating the terms of a knowledge domain, it should be checked whether there are controlled vocabularies and/or ontologies that already cover the domain or a very close domain substantially or partially. Such conceptualizations should be considered for inclusion, refining the granularity of the ontology, harmonization, and standards alignment.

If multiple ontologies define the same term, even with a slightly different name, it has to be assessed whether the represented concept or role corresponds to the same real-world object or object property. For example, while similar, there is a difference between dc:creator and foaf:maker. If the creator of a resource is declared using a string literal, dc:creator is preferred (which has neither a domain nor a range declared), whereas if the creator of a resource has a unique identifier (URI or IRI), foaf:maker is the better choice.

## 5.1.5 Enumerating the Terms of the Knowledge Domain

After assessing the use of existing ontologies, the terms required to describe the knowledge domain of the ontology should be collected. Is it important to distinguish between a concept and its name (atomic concept), because concepts represent real-world concepts of the knowledge domain of the ontology, not the words that denote them. Synonyms of the same concept do not represent different concepts, and should be declared as equivalent concepts. At this stage, properties and relationships are not important, and no structure is needed for the terms yet.

## 5.1.6 Building the Concept Hierarchy

The terms that represent objects with independent existence are selected from the set of terms collected in the previous step, and are named to create classes for the ontology. After defining the classes, they are arranged in a taxonomic (subclass-superclass) hierarchy with subsumption declarations, in which, if a class $C$ is a superclass of class $D$, then every instance of $D$ is also an instance of $C$, formally $D \sqsubseteq C$, e.g., BroadcastEvent $\sqsubseteq$ Event (every broadcast event is an event). As usual, all concepts are contained by $\top$ (when implemented in OWL, each class is a subclass of owl:Thing, and owl:Nothing is a subclass of every class). The main methods to define class hierarchies by the direction of taxonomy construction are top-down (define the generic classes first), bottom-up (define the specific classes first), and hybrid (define the known classes first) [171].

Inspired by set theory, most description logics do not have variables,[1] and provide universal statements for the TBox not only in the form of concept names

---

[1]One of the exceptions is $\mathcal{SROIQV}$.

(captured during conceptualization), but also as complex concepts consisting of concept names and Boolean operators, such as Cartoon ⊓ LiveAction ⊑ CartoonWithMotionPicture ⊔ ¬ComputerAnimation.

One of the hardest decisions of ontology design is to determine when to add a new concept or represent the differences between concepts using role values. To avoid extremes, such as overly specific classes, very low number of roles, a flat concept hierarchy, or the representation of most concept differences as role values, the following rule of thumb should be applied. If the subconcepts of a concept have roles the superconcept does not have, the role restrictions of a subconcept are different from those of the superconcept, or the subconcepts are associated with different relationships than the superconcept, new concepts have to be introduced.

While there are no rules regarding the minimum and maximum number of concepts of an ontology, there are two guidelines that can indicate problems with modeling. If a concept has only one direct subconcept, the modeling might be incorrect or incomplete. If there are numerous subconcepts for a particular concept, it is likely that additional intermediate categories are needed. When creating the concept hierarchy, care must be taken to prevent concept cycles, i.e., concept $C$ cannot be a subconcept and a superconcept of concept $D$ at the same time.

## 5.1.7   Defining Roles

After defining the concept hierarchy, the next step is defining the roles (and where applicable, their permissible values) and declaring the domain and range of the properties with the most general concept(s) possible. Role declarations include role inclusions of the form $R \sqsubseteq S$, complex role inclusions of the form $R_1 \circ R_2 \sqsubseteq S$, and role assertions of the form $Disjoint(R, S)$. Equivalent roles are declared as $R \equiv S$. The generalization-specialization relationships can typically be defined in the form of "is a" or "kind of" relationships.

For example, starredIn(JOHNWAYNE, TRUEGRIT) states that John Wayne was starred in *True Grit* (the relation between John Wayne and True Grit is "starred in"). Domains are declared in the form $\exists R.\top \sqsubseteq C$, such as $\exists surname.\top \sqsubseteq Person$, whereas range declarations are of the form $\top \sqsubseteq \forall R.D$, as, for example, $\top \sqsubseteq \forall$ releaseDate.date. Overly general domain and range declarations should be avoided.

Deciding whether a specific distinction should be modeled as a set of specific concepts or a role value can be challenging, and it depends on the domain scope and the anticipated applications. As a general rule of thumb, if the concepts with different role values mean restrictions for roles of other concepts, a new concept has to be added, if not, the distinction should be represented as a role value. Also, if a distinction is important for the represented knowledge domain, and the objects with different values are of a different kind, the distinction should be represented as a new concept rather than a role value.

## 5.1.8   Adding Individuals

Individuals can form the ABox of an ontology (knowledge base), but can also be defined outside ontologies, such as in LOD datasets. RDF statements about individuals from virtually any kind of resource can be reused in individual assertions. Individuals capture ground facts in the form of assertional knowledge. For example, `Actor(JOHNWAYNE)` indicates that the individual named `JOHNWAYNE` belongs to the `Actor` concept.

Deciding whether an object is a concept in an ontology or an individual depends on the anticipated applications, which determine the lowest level of granularity. The most specific objects should be individuals, not concepts. Also, those objects that form a natural hierarchy should be represented as concepts rather than individuals. Individual assertions not only collect information about individuals, but can also serve as examples for a particular concept type so that implementers will fully understand the intended use and specificity of the concepts defined in an ontology.

## 5.1.9   Creating a Ruleset

If the ontology language is not expressive enough to provide the intended semantics, rule languages can be used to extend the expressivity of the ontology language.

SWRL rules written in the presentation syntax can be translated to code using ontology editors such as Protégé (see later), or parsed using the OWLAPI application programming interface.[2]

Similar to DL axioms, SWRL rules can be interpreted only in terms of the ontology for which they are designed, thus they are usually defined in the same file as the DL axioms.

## 5.1.10   Evaluation

Ontologies can be evaluated according to the five ontology engineering principles of Gruber, the researcher who introduced ontologies in the context of artificial intelligence [172]:

- *Clarity*. The intended semantics of the defined terms have to be provided in a human-readable form, complemented by machine-interpretable constraints.

---

[2]https://github.com/owlcs/owlapi

- *Coherence.* Ontology axioms should be coherent so that no contradictory RDF statements can be automatically inferred from them. This should be tested using semantic reasoners.
- *Minimal encoding bias.* The conceptualization of the knowledge domain has to be specified at the knowledge level independent from any symbol-level encoding. The ontology engineering has to be conducted using open standards rather than proprietary specifications, serializations, or file formats.
- *Minimal ontological commitment.* Ontologies must be designed to be as light-weight as possible and open to more specific implementations. Vocabularies should be used consistently according to the theory specified by the corresponding ontologies.
- *Extendibility.* Ontologies should be designed in a way that new concepts, roles, and individuals can be easily added to the ontology without changing the core concept or role hierarchy.

## *5.1.11    Documentation*

The importance of ontology documentation is often overlooked. The documentation of an ontology can provide a general description of the ontology, important technical characteristics, contact persons, information about the anticipated audience and applications, as well as sample code for implementation. Online documentation should be provided for all ontologies on the dedicated website of the ontology (if there is one) or on a software repository site such as GitHub. The documentation should be updated with each version of the ontology.

The precise documentation of ontologies with the degree of formality chosen to the anticipated application can be achieved using OMDoc documents, in which symbols (that correspond to ontology entities) can be composed to formulae encoded in OpenMath or MathML [173].

## *5.1.12    Maintenance*

After being designed and published, ontologies have to be continuously maintained. Ontology maintenance is particularly important when major corrections are made, extensions are added, or a new version is released. Most knowledge domains evolve over time, making it necessary to update ontologies to reflect the current technology, new individuals, and so on.

The web hosting of the ontology files has to be renewed in due time (or the registration maintained on software repositories), and if the IRI of the ontology changes, the permalinks pointing to the ontology should be updated (e.g., on purl.org). The ontology namespace and preferred prefix should be used consistently, so that the axioms of ontologies and statements of LOD datasets that use terms from the ontology remain valid. Provenance data should also be maintained, for example, when new versions reuse additional vocabularies or ontologies.

## 5.2   Ontology Engineering Tools

When manual editing of ontology files is not a viable option or is simply inefficient, knowledge engineers can choose from a variety of powerful software tools to assist ontology development. Ontology editors are used widely in ontology engineering for creating, manipulating, and visualizing ontologies, many of which support reasoning through plugins. There are special software tools for ontology analysis and ontology mapping. Specific software libraries are available for developing ontology-based Semantic Web applications. Reasoners can be used not only for integrity checking, but also for automated scene interpretation and knowledge discovery. Rule engines are suitable for efficient information processing, combination, and analysis.

### 5.2.1   Ontology Editors

Ontology editors are software tools especially designed for ontology engineering. They cover the common tasks of the major stages of ontology development by providing an interface to do the following:

- Read and write ontology files in a variety of serialization formats
- Maintain licensing and provenance data, and ontology metadata
- Manipulate namespaces with or without versioning, and change the ontology IRI throughout an ontology from a central location
- Automatically identify core integrity issues
- Visualize the taxonomical hierarchy of concepts in a TBox
- Add and update classes, properties, and individuals
- Define, edit, and apply datatypes
- Distinguish between explicitly stated and inferred RDF statements
- Transfer user-selected axioms across ontologies
- Merge multiple ontologies into a single ontology
- Count metrics, such as the number of classes, properties, and individuals
- Calculate description logic expressivity
- Perform reasoning (typically through plugins)
- Visualize ontologies as RDF graphs
- Compare ontologies
- Query knowledge bases

### 5.2.1.1 Protégé

*Protégé*[3] is the most widely used open source ontology editor and knowledge management toolset. It supports a variety of plugins, including reasoners (e.g., HermiT, FaCT++, ELK), SWRL editors (e.g., SWRLTab), and visualization tools (e.g., VOWL, NavigOWL). Originally developed as a learning health system for translating raw biomedical data into machine-readable data for decision making, Protégé is now suitable for modeling arbitrary knowledge domains for ontology-driven applications. The online version[4] even supports collaborative ontology engineering.

The intuitive graphical user interface (GUI) of Protégé, which features a main menu, an address bar, and a tab-based editor (see Fig. 5.1), visualizes class and role hierarchy. This makes it easy to oversee the ontology being edited; however, the OWLAPI that powers Protégé overwrites any manually optimized ontology source and does not prevent ontology engineering issues, as will be discussed later in Chap. 5.

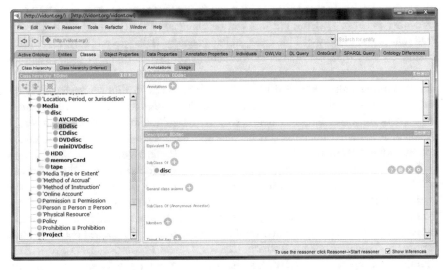

**Fig. 5.1** Protégé's GUI provides easy access to classes, object and data properties, and individuals, and visualizes taxonomical hierarchy

In the File menu, existing ontologies can be opened from an offline or online RDFS or OWL ontology file, and new ontologies can be created and saved in many formats including, but not limited to, RDF/XML, Turtle, OWL/XML, the OWL functional syntax, the Manchester syntax, and JSON-LD.

---

[3]http://protege.stanford.edu

[4]http://webprotege.stanford.edu

Under File ➤ Preferences, hidden annotation URIs can be managed. Unique identifiers can be generated automatically for classes, properties, and individuals by setting up or modifying the structure of entity URIs for the ontology being edited. Once the base URI is set up for the ontology, all fragment identifiers of the ontology will start with that URI, which can be modified any time later (New Ontologies tab in File ➤ Preferences). This is particularly useful if the address structure has to be changed for existing ontologies. A base URI is an arbitrary URI, and the path can optionally include the current year, month, and day. The base URI ends in a # by default, but this can be changed to / if needed (New Entities tab in File ➤ Preferences). The ending of the entity URIs can be set to an arbitrary name, which is the default option. To use automatically generated identifiers instead, entity labels can be set, including custom URIs and a globally unique prefix or suffix.

OWLViz, a built-in Protégé plugin, powers the graphical representation of class hierarchies of OWL ontologies and the navigation between the classes represented as a tree structure (OWLViz tab in File ➤ Preferences). OWLViz makes the comparison of the asserted class hierarchy and the inferred class hierarchy possible. By default, Protégé automatically checks for plugin updates at program startup, which can also be disabled (Plugins tab in File ➤ Preferences). The default plugin repository is set to GitHub, but this can be changed. The Reasoner tab in File ➤ Preferences can display or hide class, object property, datatype property, and object inferences, and initialize reasoners via setting up precomputation tasks, such as classification, to be performed when the reasoner is launched. The tree hierarchy can be automatically expanded under Tree Preferences in File ➤ Preferences by setting an autoexpansion depth limit (the default value is 3) and an autoexpansion child count limit (the default value is 50). By default, automatic tree expansion is disabled. Accidental changes can be reverted by clicking the Reset preferences button on the bottom right-hand corner of File ➤ Preferences.

The tabs collect the core ontology engineering functions and the different ontology views. The Active Ontology tab shows general ontology metadata, such as title, creator, description, imported ontologies, and ontology statistics, such as the number of axioms, classes, object properties, individuals, and description logic expressivity. Protégé also displays all the prefixes used in the opened ontology, and features a dedicated tab for Entities, Classes, Object Properties, Data Properties, Annotation Properties, and Individuals. The class hierarchy is shown as a tree structure, where each node can be opened or closed individually. The selected entity, class, or property details are shown in separate panels. Class descriptions provide information about equivalent classes, subclasses, class axioms, and members of the selected class. By default, the classes in Protégé are subclasses of `owl:Thing`. Class hierarchies can be created from the Tools menu. Properties can be set to subproperties and inverse properties of other properties, and be functional, transitive, symmetric, asymmetric, reflexive, or irreflexive. Protégé automatically updates the declarations of inverse properties.

The Object Properties and Data Properties tabs also have a Characteristics panel. For object properties, the Characteristics panel features checkboxes for functional, inverse functional, transitive, symmetric, asymmetric, reflexive, and irreflexive properties. The Individuals tab shows not only the class hierarchy, but also the members list and the property assertions. The OntoGraf tab provides a visual representation of the ontology being edited. When hovering the mouse over any part of the graph, Protégé displays the corresponding fragment identifier, as well as the subclasses/superclasses (if any).

The SPARQL Query tab provides an interface to execute arbitrary SPARQL queries. This interface enumerates the prefixes and provides an editable SELECT query template for easy query writing.

The ontologies created in Protégé can be accessed from Java programs through the Protégé-OWL API.

### 5.2.1.2   SemanticWorks

*SemanticWorks* is a visual Semantic Web editor that features a graphical RDF/RDFS editor and an OWL editor [174]. SemanticWorks is capable of syntax and format checking and evaluating ontology semantics with direct links to errors. Context-sensitive entry helpers display the list of valid input options for the serialization being used. SemanticWorks supports saving to RDF/XML and N-Triples, and RDF/XML to N-Triples conversion. The program features a printing option for RDF and OWL diagrams. New class instances can be defined using intelligent shortcuts. The instances, properties, and classes are organized on tabs, and, similar to software engineering environments, properties and property values can also be manipulated through separate subwindows. The Overview subwindow is very useful when editing large, complex diagrams, because the currently displayed portion of the diagram is indicated as a red rectangle. SemanticWorks allows switching between the diagram and the code view at any time.

### 5.2.1.3   TopBraid Composer

*TopBraid Composer* is a graphical development tool for data modeling and semantic data processing. The free Standard Edition supports standards such as RDF, RDFS, OWL, and SPARQL for visual editing and querying, and data conversions [175]. On the top of these features, the commercial Maestro Edition provides a model-driven application development environment [176].

TopBraid Composer can open ontologies serialized in RDF/XML or Turtle, and import RDFa data sources, RSS or Atom news feeds, and e-mails into RDF. It can connect to SPARQL endpoints as well as RDBMS sources, and import tab-delimited spreadsheet files and Excel spreadsheets, online RDF and OWL files, UML files, XML schemas, and XML catalogs. Wizards guide developers in creating new projects, such as faceted project resources, projects from CSV files, JavaScript projects, static web projects, as well as XML editing and validation. Markup files can be created using RDFa and HTML5 Microdata annotations, semantic web applications developed, and RDF/OWL files connected to Jena

SDB databases, Jena TDB databases, Oracle databases, and Sesame 2 repositories. The GUI features panels for classes, visual representation in the form of diagrams and graphs, the source code, properties, file system navigation, imports, and so-called baskets (for collecting Linked Data during browsing), as shown in Fig. 5.2.

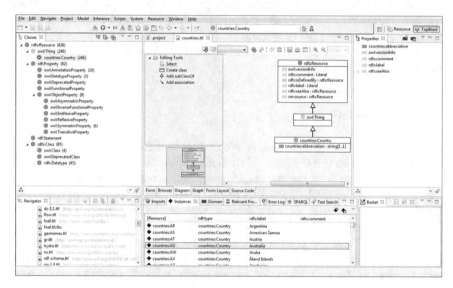

**Fig. 5.2** The GUI of TopBraid Composer Maestro displays the taxonomical hierarchy, entity diagram, property list, and ontology files

The Classes panel enables navigation in ontologies, displaying ontologies in a tree structure, creating and deleting classes, creating subclasses, grouping components by namespace, and searching by name. The Properties panel not only allows property manipulation, but also features GoogleMaps integration. On the Imports panel, the resources can be displayed, along with their rdf:type, rdfs:label, and rdfs:comment values (when provided), as well as rules, instances, errors, SPARQL queries, and text searches. On the Baskets panel, contents can be opened from, and saved to, a text file, and selected resources can be added, along with matching properties, subclasses, subproperties, instances, and individuals. Moreover, resources can be dereferenced and batch operations can be performed.

#### 5.2.1.4  Apache Stanbol

*Apache Stanbol*[5] is a semantic data modeler and comprehensive ontology manager. It includes a content management system that supports Semantic Web services and web application functions, such as tag extraction, text completion in search fields,

---

[5]http://stanbol.apache.org

and e-mail routing, based on extracted entities. The functionalities of the Stanbol components are available through a RESTful API. The RESTful services return results in RDF, JSON, and JSON-LD. Apache Stanbol can be run as a stand-alone application (packaged as a runnable JAR) or as a web application (packaged as .war) deployable in servlet containers such as Apache Tomcat. It is compatible with Apache frameworks such as Solr (for semantic search), Tika (for metadata extraction), and Jena (for storage).

Stanbol has a built-in RDFizer that processes traditional web contents sent in a POST request with the MIME type specified in the Content-type header and adds semantic enrichment (RDF enhancement) to it, serialized in the format specified in the Accept header.

Stanbol also provides a reasoner component, which implements a common API and supports different reasoners and configurations through OWLAPI and Jena-based abstract services, with implementations for Jena, RDFS, OWL, OWLMini, and HermiT. The reasoner module can perform consistency checks, which return HTTP Status 200 if the data is consistent and 204 if not. The reasoner of Stanbol is particularly powerful in classification.

The Apache Stanbol Ontology Manager enables the management of ontology networks by interconnecting seemingly unrelated knowledge represented across ontologies. It supports common ontology engineering tasks, such as reasoning and rule execution. Stanbol can also store and cache semantic information and expose it to knowledge-based information retrieval.

### 5.2.1.5  Fluent Editor

*Fluent Editor*[6] is an ontology editor, which can handle RDF, OWL, and SWRL files. The editor uses a proprietary representation language and query language compatible with Semantic Web standards. The tool has been designed for managing complex ontologies. It features a reasoner window, a SPARQL window for queries, an XML preview window, a taxonomy tree view, and an annotation window. Fluent Editor has two types of plugins: a Protégé interoperability plugin, which supports data export to, and import from, Protégé, and R language plugins that support the development of analytical models with R and rOntorion and plugin development for Fluent Editor with the R language.

---

[6]http://www.cognitum.eu/semantics/FluentEditor/

## 5.2.2  Ontology Analysis Tools

There are software tools for ontology mapping and specific ontology engineering tasks not supported by general-purpose ontology editors, such as semantic similarity estimation.

### 5.2.2.1  ZOOMA

*ZOOMA*[7] is an application for discovering optimal ontology mappings and automatic mapping of text values to ontology terms using mapping repositories. ZOOMA can reuse mappings already asserted in the database, explore mapping best suitable for multiple mappings, derive better mappings by recording contextual information, and suggest new terms. The commonly observed values can be processed automatically.

ZOOMA finds all optimal mappings automatically where one text value maps to the same set of terms every time. When using mapping repositories, it can detect errors; in other words, it finds all the text values to ontology term mappings that are potentially incorrect. ZOOMA can also propose new mappings to terms based on the input values; however, selecting the best mapping requires human evaluation and assessment. ZOOMA can easily be used as a software library, as, for example, within an Apache Maven project.

### 5.2.2.2  Semantic Measures Library

The *Semantic Measures Library (SML)*[8] is a Java library for semantic measure analysis, such as estimating semantic similarity and relatedness by using ontologies to define the distance between terms or concepts. SML functionalities can also be accessed through a set of command-line tools called SML-Toolkit. The library supports RDF and RDFS, OWL ontologies, WordNet, Medical Subject Headings (MeSH, a controlled vocabulary for life science publishing), the Gene Ontology, and so on.

## 5.3  The Evolution of Multimedia Ontology Engineering

While a large number of structured multimedia ontologies have been created over the years, many are limited by various design issues. For example, the well-known *Large Scale Concept Ontology for Multimedia (LSCOM)* defines atomic concepts

---

[7]http://www.ebi.ac.uk/fgpt/zooma/

[8]http://www.semantic-measures-library.org

and their hierarchical structure, so it is a taxonomy extended with some basic ontology constructors (atomic negation, concept intersection, universal restrictions, and limited existential quantification, which correspond to the basic Attributive Language, $\mathcal{AL}$) [177]. It is not a fully featured ontology, because it does not axiomatize the sophisticated relationships between concepts and roles, and does not define any complex role inclusions. Common concepts of everyday life depicted in images and videos can also be described using common sense ontologies, so the purpose of the LSCOM ontology is somewhat disputable.[9] The *Visual Descriptor Ontology (VDO)*, which was published as an "ontology for multimedia reasoning" [178], is actually unsuitable for reasoning over multimedia contents, because it has a very limited description logic expressivity (corresponding to $\mathcal{ALH}$), and lacks the formalized description of complex relationships between concepts and roles. A list of terms lacking a comprehensive formal description is unquestionably insufficient for multimedia reasoning. While finding the right trade-off between expressivity and computational complexity is crucial in ontology engineering, extremely low expressivity might prevent reasoning altogether. The *Core Ontology for Multimedia (COMM)* uses $\mathcal{SHOIN}^{(\mathcal{D})}$, the most expressive DL at the time of its release (in OWL DL); however, the current standard, OWL 2 DL, supports far more constructors, such as complex role inclusion axioms; reflexive, asymmetric, irreflexive, and disjoint roles; universal role; self-constructs; negated role assertions; and qualified number restrictions. At the time of writing, the only multimedia ontologies that exploit the full range of OWL 2 constructors ($\mathcal{SROIQ}^{(\mathcal{D})}$ DL) are *VidOnt*[10] and the *3D Modeling Ontology*.[11]

The following sections give an overview of the bad practices of multimedia ontology engineering, and confront them with best practices to demonstrate the importance of proper ontology design and formal grounding in description logics.

## 5.3.1   Semistructured Vocabularies

Because of the potential of DL mapping, the first controlled vocabularies and thesauri of interest are the semistructured, XML- or XSD-based vocabularies and thesauri (see Table 5.1).

---

[9]In 2006, the LSCOM terms were mapped to Cyc terms and integrated into ResearchCyc, which indicates that the ontology could be part of a common sense ontology. However, LSCOM itself is not a common sense ontology, and it is not a domain ontology either. The concept categories of the 2006 TRECVID dataset, on which the multimedia concepts of LSCOM are based, correspond to multiple knowledge domains. While the LSCOM concepts are commonly depicted concepts in particular domains, the concept coverage of the ontology is not sufficient for annotating arbitrary videos. For very specific domains, LSCOM would provide too generic concepts. LSCOM is not suitable for the spatiotemporal annotation of videos and video RoIs or multimedia reasoning either. For annotating commonly depicted atomic concepts, common sense or upper ontologies, and for specific concepts a domain ontology, would provide far better concept coverage.

[10]http://vidont.org/vidont.ttl

[11]http://3dontology.org/3d.ttl

**Table 5.1** Semistructured multimedia vocabularies and thesauri

| Vocabulary/ontology | Standard | Language |
|---|---|---|
| EBU Genre Vocabulary | De facto standard | XML[a] |
| Media Contract Ontology (MCO) | ISO/IEC 21000 (MPEG-21) | XSD[b] |
| Multimedia Content Description Interface (MCDI) | ISO/IEC 15938 (MPEG-7) | XSD[c] |
| Thesaurus for Graphic Materials (TGM I, TGM II) | – | XML[d] |
| TV-Anytime | IETF RFC 4078[e] | XSD[f] |

[a]https://www.ebu.ch/metadata/cs/ebu_ContentGenreCS.xml
[b]http://purl.org/NET/mco-core, http://purl.org/NET/mco-ipre
[c]https://www.iso.org/standard/34230.html
[d]https://www.loc.gov/rr/print/tgm1/tgm1.xml, https://www.loc.gov/rr/print/tgm2/tgm2.xml
[e]https://tools.ietf.org/html/rfc4078
[f]http://webapp.etsi.org/workprogram/Report_WorkItem.asp?WKI_ID=39864

For those vocabularies that have been created before the Semantic Web era, such as TGM I, which appeared in 1980 [179], using semistructured definitions became a straightforward choice with the introduction of XML in 1996. In addition, constraints can be added to the terms using XML Schema, for example, to declare the range of permissible values for roles. XML-based vocabularies can be translated to RDFS controlled vocabularies, which not only describe a list of words, but also identify synonyms, disambiguate homographs, and identify relationships between terms, as discussed earlier in Chap. 2. Mapping the vocabulary of standards such as MPEG-7, MPEG-21, and TV-Anytime to OWL could be useful for defining richer semantics for, among others, complex roles. However, most of the mappings released to date provide nothing more than the OWL equivalent of the original XML-based definitions, sometimes extended with some basic constructors, as you will see in the next section.

### 5.3.2  Structured Ontologies Mapped from Semistructured Vocabularies

Many attempts have been made to map the XML-based low-level descriptors of MPEG-7 to RDFS and OWL. Similarly, MPEG-21 and TV-Anytime have also been mapped to OWL (see Table 5.2).

**Table 5.2** Standard-based multimedia ontologies

| Ontology | Base standard | Language | DL expressivity |
|---|---|---|---|
| Content Ontology for the TV-Anytime Content CS | TV-Anytime Content | OWL[a] | $\mathcal{AL}$ |
| Core Ontology for Multimedia (COMM) | MPEG-7 Structure and Visual | OWL[b] | $\mathcal{SHOIN}^{(D)}$ |
| Format Ontology for the TV-Anytime Format CS | TV-Anytime Format | OWL[c] | $\mathcal{AL}$ |
| Media Value Chain Ontology (MVCO) | ISO/IEC 21000-19:2010 (MPEG-21)[d] | OWL[e] | $\mathcal{SIN}^{(D)}$ |
| Rhizomik | MPEG-7 | OWL[f] | $\mathcal{ALUN}^{(D)}$ |
| SmartWeb Integrated Ontology (SWIntO) | MPEG-7 Structure and Visual | RDFS[g] | $\mathcal{ALH}^{(D)}$ |
| Visual Descriptor Ontology (VDO) | MPEG-7 Visual (with revised structure) | OWL[h] | $\mathcal{ALH}$ |

[a] http://rhizomik.net/ontologies/2005/03/TVAnytimeContent.owl
[b] http://multimedia.semanticweb.org/COMM/visual.owl
[c] http://rhizomik.net/ontologies/2005/03/TVAnytimeFormat.owl
[d] https://www.iso.org/standard/52887.html
[e] http://purl.oclc.org/NET/mvco.owl
[f] http://rhizomik.net/ontologies/2005/03/Mpeg7-2001.owl
[g] http://smartweb.dfki.de/ontology/swinto0.3.1.rdfs
[h] https://raw.githubusercontent.com/gatemezing/MMOntologies/master/ACEMEDIA/acemedia-visual-descriptor-ontology-v09.rdfs.owl

While the MPEG-7 mappings provide structured data (as opposed to the original semistructured data), they inherited several issues of the original definition of the MPEG-7 standard, as discussed earlier in Chap. 3. As shown above, the Core Ontology for Multimedia is the only MPEG-7-based ontology that exploits all the mathematical constructors of the underlying description logic. All the other "ontologies" are far less expressive and barely define more than a list of terms with a class hierarchy, therefore most of them are controlled vocabularies, taxonomies, or thesauri only. In addition, they have multiple design issues. Firstly, the potential application areas of these ontologies have not been clearly defined, which should have been done in the first step of ontology engineering. Secondly, their conceptualization is usually poor, which indicates the lack of the conceptual model and that they have either inadequate formal grounding or no grounding at all.[12] Although these mapped ontologies transformed semistructured definitions to structured data, they are still not suitable for reasoning over visual content. Since MPEG-7 provides low-level descriptors, their OWL mapping does not provide real-world semantics, which can be achieved through high-level descriptors only. The MPEG-7 descriptors provide metadata and technical characteristics to be processed by computers, so

---

[12] It is a common but bad practice to create OWL ontologies using a tree structure of concept hierarchies visualized by the GUI of Protégé without formal grounding in description logics.

their structured definition does not contribute to the semantic enrichment of the corresponding multimedia resources. To demonstrate this, take a closer look at the code of COMM (see Listing 5.1).

**Listing 5.1** Code fragment of COMM

```
<owl:Class rdf:about="#cbac-coefficient-20">
  <rdfs:comment rdf:datatype="&xsd;string">Corrresponds to
  the &quot;CrACCoeff20&#8221; element of the
  &quot;ColorLayoutType&quot; (part 3, page 45-46)
  </rdfs:comment>
  <rdfs:subClassOf>
    <owl:Class rdf:about="&p1;unsigned-5-vector-dim-20"/>
  </rdfs:subClassOf>
  <rdfs:subClassOf>
  <owl:Class rdf:about="#cbac-crac-coefficient-20-descriptor-
  parameter"/>
  </rdfs:subClassOf>
</owl:Class>
```

This example is related to the *color layout descriptor (CLD)* of MPEG-7 mentioned earlier in Chap. 3, which is designed to capture the spatial distribution of color in images for fast browsing and retrieval. The algorithm that computes the CLD partitions an image (RGB typically converted to the YCbCr color space) into $8 \times 8$ subimages,[13] and calculates the dominant color of each subimage. Then it performs the *discrete cosine transform (DCT)* of this $8 \times 8$ matrix of dominant colors to the luminance (Y) and the blue and red chrominance (Cb and Cr), which yields to three sets of 64 signal amplitudes, i.e., DCT coefficients (DCTY, DCTCb, and DCTCr). The DCT coefficients include DC coefficients and AC coefficients,[14] i.e., coefficients that have zero frequency in both dimensions (the mean value of the waveform) and coefficients that have nonzero frequencies. These DCT coefficients are then quantized and finally zig-zag scanned.

Hence, the cbac-coefficient-20 in Listing 5.1 can be used to represent the blue chrominance AC coefficients, which are calculated by computer software to generate the CLD for an image or video frame. The color layout descriptor is suitable for content-based image retrieval and filtering, and the computer programs that generate and use the color layout descriptor can process the descriptor data even if it is written in XML (using terms from the vocabulary of the original MPEG-7 standard). Therefore, while the XML to OWL translation is straightforward, such as via XSLT (e.g., XML2OWL),[15] the OWL mapping of such low-level

---

[13]These 64 blocks guarantee resolution invariance and scale invariance.

[14]The names AC and DC are derived from the historical use of the DCT for analysing electrical currents (AC—alternating current, DC—direct current).

[15]http://xml2owl.sourceforge.net

descriptors is not particularly useful. You see, the real power of OWL in creating multimedia ontologies lies in the formal representation of the depicted knowledge domain, which can provide rich semantics of an image or video for humans (such as during hypervideo playback), and enables advanced tasks for computers via reasoning. Since the OWL mappings of MPEG-7 have been created neither for humans nor for computers, the anticipated application area was clearly not defined properly.

Even the Core Ontology for Multimedia, which is one of the most sophisticated OWL mappings of MPEG-7, defines the document of the MPEG-7 specification itself rather than actual multimedia components. It defines the above `cbac-coefficient-20` class as a subclass of `cbac-crac-coefficient-20-descriptor-parameter` and `unsigned-5-vector-dim-20`. The `unsigned-5-vector-dim-5` class is a subclass of `integer-vector`, which is a subclass of `vector`, which is a subclass of `abstract-region`, which is a subclass of `region`, which is a subclass of `particular`, which is a subclass of `owl:Thing`. By formalizing these classes in description logics, the ontology design issues become obvious:

cbac-coefficient-20 $\sqsubseteq$ unsigned-5-vector-dim-20
unsigned-5-vector-dim-20 $\sqsubseteq$ integer-vector
integer-vector $\sqsubseteq$ vector
vector $\sqsubseteq$ abstract-region
abstract-region $\sqsubseteq$ region
region $\sqsubseteq$ particular
particular $\sqsubseteq$ $\top$

The ontology does not even conceptualize proper superclasses, such as image, audio, and video, and the uppermost user-defined concept is `particular`. In other words, the knowledge domain of this ontology is not the multimedia domain (as it should have been), but the domain of multimedia terms from the MPEG-7 vocabulary. If the knowledge domain had been correctly determined during ontology engineering, the coefficients would have been defined as roles and not concepts, so that they could have been used to hold the corresponding values as role values.

From the code optimality point of view, the file of the COMM ontology is full of whitespace characters, and the description of the subject resource provided using `rdfs:comment` employs entity codes rather than directly written UTF-8 characters. From the assessment and maintenance points of view, the typo in the description, which occurs at multiple locations in the code, should have been found and corrected.

### 5.3.3    *Structured Multimedia Ontologies*

Structured multimedia ontologies, many of which are intended to cover certain aspects of multimedia resources, are available for both audio and video resources. However, the knowledge domain and scope of these multimedia ontologies are often not clearly defined, and many multimedia ontologies incorrectly attempt to cover domains such as multimedia, multimedia resources, terms defined by the MPEG-7 standard, and concepts commonly depicted in images and videos, instead of more reasonable knowledge domains for multimedia resources, such as images, audio contents, animations, videos, and 3D models. Quite often, multimedia ontologies are so poorly designed that even their type is unclear: many of them do not fall into any known ontology category.

While reusing terms from standardized ontologies is a golden rule on the Semantic Web, potential term reuse is often not assessed properly, and many already defined terms are redefined by multimedia ontologies. If there are very specific definitions already available, defining overly general new terms is unacceptable. For example, there are more specific definitions for many of the atomic concepts provided by the Linked Movie Database ontology in Dublin Core, Creative Commons, DBpedia, and VidOnt, just to name a few. Moreover, a movie ontology should incorporate relevant audio and video terminology as well, not only movie features.

Conceptualization issues usually derive from a lack of properly determined domain and scope. For example, while the Linked Movie Database was designed especially for movies, it only covers very general terms for film distribution, not professional filmmaking. This indicates that the scope was poorly defined, omitting important concepts associated with Hollywood movies, such as trailers and video clips.

Poor conceptualization affects concept inclusion and the creation of the concept hierarchy, although most multimedia ontologies feature a reasonably designed hierarchy for their concepts, mainly because creating the taxonomical structure is the main focus of ontology engineers. However, most multimedia ontologies do not define complex roles and rules, and are not detailed enough to capture the role constraints, such as domain, range, and permissible values, clearly indicated by the low DL expressivity of most multimedia ontologies. Even well-established structured multimedia ontologies, most of which are domain ontologies, leverage the most basic mathematical constructors only, as many of them correspond to the basic $\mathcal{AL}$ description logic (see Table 5.3). While this certainly makes them lightweight, it also limits their application potential.

**Table 5.3** Core vocabularies and ontologies for multimedia applications

| Vocabulary/ontology | Standard | Language | DL expressivity |
|---|---|---|---|
| 3D Modeling Ontology (3DMO) | X3D-aligned | OWL 2[a] | $\mathcal{SROIQ}^{(\mathcal{D})}$ |
| Audio Effects Ontology (AUFX-O) | – | OWL[b] | $\mathcal{ALCRIF}^{(\mathcal{D})}$ |
| BBC Programmes Ontology | – | OWL[c] | $\mathcal{SHOIN}^{(\mathcal{D})}$ |
| Chord Ontology | – | RDFS[d] | $\mathcal{ALU}^{(\mathcal{D})}$ |
| EBU CCDM | De facto standard | OWL[e] | $\mathcal{ALCI}^{(\mathcal{D})}$ |
| EBU Core | De facto standard | OWL[f] | $\mathcal{ALCI}^{(\mathcal{D})}$ |
| IPTC NewsCodes | De facto standard | RDFS[g] | $\mathcal{AL}$ |
| Keys Ontology | – | OWL[h] | $\mathcal{AL}^{(\mathcal{D})}$ |
| Large Scale Concept Ontology for Multimedia (LSCOM) | – | OWL[i] | $\mathcal{AL}$ |
| Linked Movie Database (LMD) | – | RDFS[j] | $\mathcal{AL}$ |
| LinkedTV | – | OWL[k] | $\mathcal{AL}^{(\mathcal{D})}$ |
| M3O: the Multimedia Metadata Ontology | – | OWL[l] | $\mathcal{SHIQ}^{(\mathcal{D})}$ |
| Multitrack Ontology | – | OWL[m] | $\mathcal{ALI}^{(\mathcal{D})}$ |
| Music Vocabulary | – | OWL[n] | $\mathcal{ALEHI}^{(\mathcal{D})}$ |
| Ontology for Media Resources 1.0 | W3C recommendation | OWL[o] | $\mathcal{ALCHI}^{(\mathcal{D})}$ |
| Video Ontology (VidOnt) | MPEG-7-aligned | OWL 2[p] | $\mathcal{SROIQ}^{(\mathcal{D})}$ |
| YOVISTO Academic Video Search Ontology | – | OWL[q] | $\mathcal{AL}$ |

[a]http://3dontology.org/3d.ttl
[b]https://w3id.org/aufx/ontology/1.0
[c]http://www.bbc.co.uk/ontologies/po/1.1.ttl
[d]http://motools.sourceforge.net/chord/index.rdfs.rdf
[e]https://www.ebu.ch/metadata/ontologies/ebuccdm/20120915/CCDM_Core.owl
[f]https://www.ebu.ch/metadata/ontologies/ebucore/20150122/ebucore_2015_01_22.rdf
[g]http://cv.iptc.org/newscodes/
[h]http://motools.sourceforge.net/keys/keys.owl
[i]http://lscom.linkeddata.es/LSCOM.owl
[j]http://data.linkedmdb.org/all
[k]http://data.linkedtv.eu/ontologies/core/core.owl
[l]http://m3o.semantic-multimedia.org/ontology/2009/09/16/annotation.owl
[m]http://purl.org/ontology/studio/multitrack
[n]http://www.kanzaki.com/ns/music
[o]https://www.w3.org/TR/mediaont-10/
[p]http://vidont.org/vidont.ttl
[q]http://www.yovisto.com

The above issues, together with the unconventional capitalization of concepts and roles (that violate description logic best practices of using PascalCase for atomic concepts and camelCase for atomic roles, as discussed earlier in Chap. 4), indicate the lack of formal grounding in description logics for most of these ontologies.

The documentation of multimedia ontologies is not always adequate; some ontologies do not have a dedicated website, and their namespace URIs are symbolic links only, while others provide a markup document for browsers and OWL for semantic agents for all their namespace URIs, either on separate web pages, or with fragment identifiers within the same web page.

The general-purpose vocabularies and ontologies used for multimedia representation include standards, such as Dublin Core, common sense ontologies, such as OpenCyc and WordNet, and upper ontologies, such as SUMO (see Table 5.4).

**Table 5.4**  General multimedia-enabled controlled vocabularies and ontologies

| Vocabulary/ontology | Standard | Language | DL expressivity |
|---|---|---|---|
| Creative Commons | De facto standard | RDFS[a] | $\mathcal{ALO}^{(\mathcal{D})}$ |
| Dublin Core | IETF RFC 5013, ISO 15836-2009, NISO Z39.85 | RDFS[b] | $\mathcal{ALH}^{(\mathcal{D})}$ |
| Event Ontology | – | OWL[c] | $\mathcal{ALCHI}^{(\mathcal{D})}$ |
| FOAF (Friend of a Friend) | De facto standard | OWL[d] | $\mathcal{ALCHIF}^{(\mathcal{D})}$ |
| OpenCyc | De facto standard | OWL[e] | $\mathcal{SH}$ |
| Schema.org | De facto standard | OWL[f] | $\mathcal{ALH}^{(\mathcal{D})}$ |
| Suggested Upper Merged Ontology (SUMO) | – | OWL[g] (translated from SUO-KIF)[h] | $\mathcal{SOIF}$ |
| Timeline Ontology | – | OWL[i] | $\mathcal{ALCHON}^{(\mathcal{D})}$ |
| WordNet 3.x | De facto standard | OWL[j] | $\mathcal{SHOIN}^{(\mathcal{D})}$ |

[a]https://creativecommons.org/schema.rdf
[b]http://dublincore.org/2012/06/14/dcterms.rdf
[c]http://motools.sf.net/event/event.n3
[d]http://xmlns.com/foaf/0.1/
[e]https://sourceforge.net/projects/texai/files/open-cyc-rdf/1.1/open-cyc.rdf.ZIP/download
[f]http://schema.org/docs/schemaorg.owl
[g]http://www.adampease.org/OP/SUMO.owl
[h]The Standard Upper Ontology Knowledge Interchange Format (SUO-KIF) is the proprietary format of the original version of SUMO
[i]http://motools.sf.net/timeline/timeline.n3
[j]http://wordnet-rdf.princeton.edu/ontology

Because most multimedia ontologies do not exploit OWL 2 constructors, and most of them not even all the OWL constructors, their application potential is limited in the analysis, understanding, and content-based retrieval of audio and video contents, animations, and 3D models, and most of them are unsuitable for spatiotemporal reasoning.

## 5.4 Ontology-Based Multimedia Annotation Tools

The following sections summarize those software tools that leverage controlled vocabularies and ontologies for manual, collaborative, or semisupervised annotation in multimedia cataloging, retrieval, and hypervideo playback.

### 5.4.1 Structured Image Annotation Tools

The well-known but discontinued *M-Ontomat Annotizer*[16] was designed to extract low-level features from images and videos, and annotate them using MPEG-7 visual descriptors, which can be represented as ontology class prototypes in RDFS for automated multimedia annotations [180]. *AKTive Media*[17] uses user-specified domain ontologies for the manual and automated annotation of multimedia content, including spatial annotation. Based on the annotation, it also attempts to retrieve related information through online queries. The *K-Space Annotation Tool (KAT)*[18] provides semiautomatic, semantic annotation of multimedia content using the Multimedia Metadata Ontology (M3O). *Pundit*[19] is a web annotation tool for the collaborative semantic annotation of online texts and images, and can be used to create RDF knowledge graphs about the annotated online resources [181]. *ImageSnippets*[20] is a collaborative image annotation tool, which generates five-star Linked Data. *Rapid image annotation with snakes (Ratsnake)*[21] is an ontology-based image and image sequence annotation tool [182], which features semiautomatic annotation, including spatial fragmentation with polygons and grids. Figure 5.3 shows a medical image annotation with Ratsnake.

However, the software saves the regions of interest to the RDF/XML output as strings ($\texttt{xsd:string}$) rather than vectors (e.g., $\texttt{xsd:complexType}$), so the coordinates of the spatial fragmentation cannot be directly manipulated. Ratsnake can be trained using reference images for each knowledge domain with presegmented and labeled objects, based on which the software can automatically annotate segments of similar images with semantic labels.

---

[16]http://mklab.iti.gr/m-onto2

[17]https://sourceforge.net/projects/aktivemedia/

[18]https://launchpad.net/kat

[19]http://thepund.it

[20]http://www.imagesnippets.com

[21]http://is-innovation.eu/ratsnake/

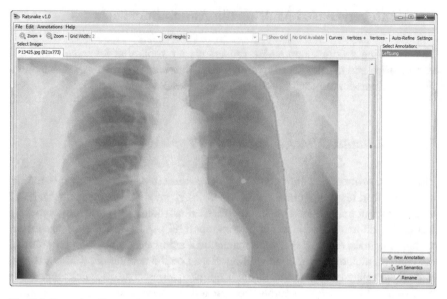

**Fig. 5.3** Ratsnake allows polygon- and grid-based spatial annotation

## 5.4.2  *Structured Audio Annotation Tools*

The *Sonic Annotator Web Application*[22] analyzes arbitrary audio files using the Vamp feature extractor plugins to estimate the tempo, beat locations, note pitch, and key of the music; detect and count zero crossing points; calculate a series of MFCC vectors from the audio; detect and return the positions of harmonic changes such as chord boundaries; and so on. Based on the extracted features, the Sonic Annotator Web Application generates RDF output.

The Music Ontology Python library (*mopy*)[23] provides Python bindings for Music Ontology terms to create and manipulate Music Ontology data.

The semantic audio annotation tool of Restagno et al. is an ontology-driven software prototype integrated with a web-based audio sequencer, which can be used for mixing and annotating sounds in a production environment [183]. The tool allows the annotation of resources with terms of user-defined ontologies in RDF. Beyond its purpose-built ontology, the *Sound Producing Events Ontology*, the tool utilizes the Music Ontology and retrieves facts from LOD datasets, such as DBpedia, via SPARQL queries for the semantic enrichment of music resources.

---

[22]https://github.com/motools/sawa

[23]https://github.com/motools/mopy

### 5.4.3  Structured Video Annotation Tools

Although image sequences, video thumbnails, and video frames could be annotated using semantic image annotation tools (as still images), the semantic annotation of videos relies on the representation of temporal information and unique video features.[24] Hence, video annotation tools not or not solely based on frame-level annotation have also been introduced [184]. First, in the late 1990s, unstructured video annotation tools appeared, such as *OVID* [185] and *Vane* [186]. They were followed by semistructured video annotation tools, such as *MuViNo*,[25] *EXMARaLDA*,[26] the *VideoAnnEx Annotation Tool*,[27] *ELAN*,[28] the *Video Image Annotation Tool (VIA)*,[29] the *Semantic Video Annotation Suite (SVAS)*,[30] *VAnalyzer*,[31] the *Semantic Video Content Annotation Tool (SVCAT)*,[32] *Anvil*,[33] and the video annotation tool of Aydınlılar and Yazıcı [187]. Structured video annotation tools, which provide RDF output, seamless ontology integration, rich semantics, and many even LOD support, are described in more detail below.

The *TV Metadata Generator*,[34] released in 2011 by Eurecom as part of the LinkedTV project, automatically converts unstructured video data, such as local and online video file metadata of TV contents, standard SRT subtitles, TV-Anytime metadata, and EXMARaLDA metadata, into RDF. However, the TV Metadata Generator cannot generate structured data based on the video content. The *LinkedTV Editor*[35] utilizes automatically generated LinkedTV annotations for broadcasting services by generating contextual information.

*Open Video Annotation*[36] is an HTML5 hypervideo application, which implements open source JavaScript libraries, including Video.js,[37] Annotator,[38] and RangeSlider,[39] to tag videos and video segments, and play semantically enriched

---

[24]There are also cross-media annotation tools, such as IMAS, YUMA, and the Video Image Annotation Tool, which provide annotations for multiple media types.

[25]http://vitooki.sourceforge.net/components/muvino/code/index.html

[26]http://www.exmaralda.org/en/tool/exmaralda/

[27]http://www.research.ibm.com/VideoAnnEx/

[28]https://tla.mpi.nl/tools/tla-tools/elan/

[29]https://sourceforge.net/projects/via-tool/

[30]https://www.joanneum.at/en/digital/productssolutions/sematic-video-annotation.html

[31]https://www.dimis.fim.uni-passau.de/iris/index.php?view=vanalyzer

[32]https://www.dimis.fim.uni-passau.de/MDPS/de/mitglieder/30-german-articles/forschung/projekte/33-svcat.html

[33]http://www.anvil-software.org

[34]https://github.com/jluisred/TVRDFizator

[35]http://editortoolv2.linkedtv.eu/

[36]https://gteavirtual.org/ova/

[37]http://videojs.com

[38]http://annotatorjs.org

[39]https://github.com/andreruffert/rangeslider.js

videos. During playback, the application can display the semantics with the video. In Open Video Annotation, video playback can also be controlled by the user via utilizing the semantic enrichment of videos, such as by selecting a video segment by its semantics (see Fig. 5.4).

**Fig. 5.4** In Open Video Annotation, notes can be added on the timeline, existing annotations viewed, and annotated video fragments played individually

Open Video Annotation is powered by the Open Annotation Data Model of the W3C, which defines the Open Annotation Ontology,[40] and implements Dublin Core, FOAF, PROV-O, rdfV, rdfsV, Representing Content in RDF,[41] SKOS, and TriG Named Graphs.[42]

*MyStoryPlayer* is a hypervideo player that supports multi-angle video annotation. In MyStoryPlayer, annotations are based on action, gesture, and posture analysis to provide RDF representation for the relationships between depicted entities [188]. The software provides general and technical video metadata, such as title and duration, as well as timestamp-based annotation of arbitrary video events and human dialogues. This makes MyStoryPlayer particularly useful in annotating educational videos and videos of performing arts (see Fig. 5.5).

*SemVidLOD*[43] is a software prototype that provides spatiotemporal annotation for online video resources, video files, and streaming media with high-level descriptors using LOD terms. SemVidLOD implements VidOnt, the most expressive multimedia ontology to date [189], to describe sophisticated high-level video contents in RDF, such as two movie characters having a conversation or crashing of a car.

The primary vocabularies and ontologies used for concept mapping vary greatly among structured video annotation tools, and include ontologies such as Dublin Core, the Ontology for Media Resources, FOAF, Open Annotation, and

---

[40]https://www.w3.org/ns/oa#

[41]https://www.w3.org/2011/content#

[42]https://www.w3.org/2004/03/trix/rdfg-1/

[43]http://vidont.org/semvidlod/

## MYSTORYPLAYER

**Fig. 5.5** The indexable classifications and timestamp-based snapshot comments of MyStory-Player are particularly useful for conversations, theatrical performances, concerts, and educational videos

Representing Content in RDF. While the support for arbitrary ontologies is highly desired, many structured video annotation tools are limited to a short list of vocabularies and ontologies for concept mapping (see Table 5.5).

**Table 5.5** Ontology use of state-of-the-art semantic video annotation tools

| Tool | Primary vocabularies and ontologies | Arbitrary ontology |
|---|---|---|
| LinkedTV Editor | Annotation Ontology, Dublin Core, FOAF, LinkedTV, Ontology for Media Resources, Open Annotation, Representing Content in RDF | – |
| MyStoryPlayer | Annotation Ontology, Dublin Core, FOAF, Open Annotation, Representing Content in RDF | – |
| Open Video Annotation | Annotation Ontology, Dublin Core, FOAF, Open Annotation, Representing Content in RDF | – |
| SemVidLOD | Dublin Core, FOAF, Schema, VidOnt | + |
| TV Metadata Generator | Annotation Ontology, Dublin Core, FOAF, LinkedTV, Ontology for Media Resources, Open Annotation, Representing Content in RDF | – |

### *5.4.4   Structured 3D Model Annotation Tools*

An early implementation of the X3D standard appeared in a semantic annotation tool that utilized an RDFS mapping of the X3D vocabulary and the `MetadataSet` node of X3D files to capture semantic descriptors of objects in 3D scenes [190]. The *3D SEmantics Annotation Model* (3DSEAM) combined MPEG-7 and X3D, and extended their vocabularies with `3DObjectType`, `3DLocator`, and `3DRegion-LocatorType`, to annotate 3D models and scenes [191]. The interactive 3D museum exhibition of the Arrigo project featured 3D models of statues, whose spherical regions of interest have been semantically annotated [192].

Some research efforts are based on the *METS Schema*[44] of the Library of Congress, with different extensions (e.g., spheres, axis-aligned boxes, and sets of vertices and surface triangles [193]), `AreaID` and `ExtMeshAreaID` attributes for `mets:area`, and new attribute values for `mets:shape`, such as `collada`, `relrect`, `relcrcle`, `relpoly`, and `extmesh` [194].

The 3D model annotation tool of Yu and Hunter utilizes the standard X3D vocabulary and domain-specific ontologies, and was specially designed for annotating 3D models of cultural artifacts [195] (see Fig. 5.6).

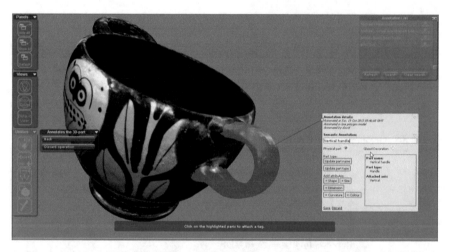

**Fig. 5.6**  Structured 3D annotation tools "understand" the geometry of 3D space, and can annotate objects and spatial fragments of objects with rich semantics. 3DSA screenshot (https://youtu.be/5qnL088qQKI)

The annotations utilize terms from the X3D vocabulary and other ontologies, such as Dublin Core, for semantic description of 3D objects and scenes. The tool implements HTML5, jQuery, and WebGL, and it is based on the Open Annotation model of the World Wide Web Consortium. The tool can capture, process, and

---

[44]https://www.loc.gov/standards/mets/

compare 3D fragment identifiers not only on low-poly, but also on high-poly meshes with up to hundreds of thousands of polygons.

## 5.5 Summary

This chapter has detailed the core steps of multimedia ontology engineering. Common mistakes and limitations have been highlighted and best practices presented. The most frequently used ontology engineering tools have been described and the most popular multimedia ontologies evaluated from the expressivity and formal grounding points of view. The thorough analysis of well-established multimedia ontologies taught a lesson about the importance of formal grounding and choosing and defining the domain and scope. To demonstrate the implementation of logic-based formalisms, the most important structured multimedia annotation tools have been listed.

# Chapter 6
# Ontology-Based Multimedia Reasoning

Description logics enable the formal representation of human knowledge about a particular domain in the form of concepts and roles, upon which implicit statements can be automatically inferred. A wide range of inference tasks (reasoning services) has been defined and investigated for a variety of description logics of different expressivity. In fact, some less expressive description logics have been introduced for the purpose of providing the desired balance of relatively good expressivity and low reasoning complexity. Even so, very expressive description logics are also in use, many of which come with PSPACE-complete, EXPTIME-complete, or even higher reasoning complexity. Despite such high complexity, DL reasoners based on the tableau method and a wide range of optimizations usually have an excellent performance in practical applications, mainly due to the relatively rare occurrence of high worst-case complexity (the average case is usually not the worst case). In the multimedia domain, reasoning is suitable for advanced classification and indexing, customized data querying, and automated scene interpretation.

## 6.1 Rationale

The use of explicitly stated information (asserted facts) alone does not exploit the full potential of description logic-based ontologies. Deriving implicit, useful information via reasoning is essential for performing advanced tasks, such as automated image or video scene interpretation, indexing, querying, and understanding 3D models. For example, the structured annotation of 3D objects enables intelligent tasks that would not be possible without semantics, such as measuring the distance between two points, material thickness, and so on. Based on these measurements, dimensional relationships can be inferred via reasoning, which is suitable, for example, to compare the volume of artifacts in a virtual museum [196] (see Fig. 6.1).

© Springer International Publishing AG 2017
L.F. Sikos, *Description Logics in Multimedia Reasoning*,
DOI 10.1007/978-3-319-54066-5_6

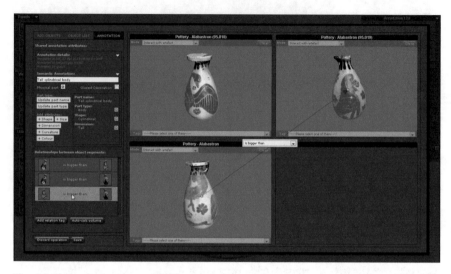

**Fig. 6.1** Reasoning over semantically enriched 3D models can be used for automatically comparing physical characteristics of the represented objects. 3DSA screenshot (https://youtu.be/5qnL088qQKI)

To enable such tasks, however, certain prerequisites must be met. The quality of the ontology has to be verified by checking the integrity of classes and instances, the class hierarchy, and the axioms. Ontology-based applications and services require high-quality ontologies that are meaningful (all concepts can have instances), correct (no contradictions), minimally redundant (no unnecessary synonyms or duplications are present), and richly axiomatized (the descriptions are sufficiently detailed). These characteristics can only be ensured via reasoning. The core reasoning tasks are described in the following sections.

## 6.2   Core Reasoning Tasks

Description logic-based knowledge representations not only store human knowledge in a machine-readable form, but also provide the option to automatically infer new RDF statements via *reasoning*, check whether the knowledge represented in a knowledge base is meaningful by searching for contradictory statements (*KB consistency*—Definition 6.1), and determine *concept satisfiability*, i.e., check whether a concept can ever have instances (see Definition 6.2). These tasks are usually performed with reference to a knowledge base $\mathcal{K} = (\mathcal{T}, \mathcal{A}, \mathcal{R})$ or a TBox $\mathcal{T}$.

**Definition 6.1 (Knowledge Base Consistency)**   Given knowledge base $\mathcal{K}$ as input, a decision procedure for knowledge base consistency returns "$\mathcal{K}$ is consistent" if there is an interpretation $\mathcal{I}$ such that $\mathcal{I} \vDash \mathcal{K}$, otherwise it returns "$\mathcal{K}$ is inconsistent."

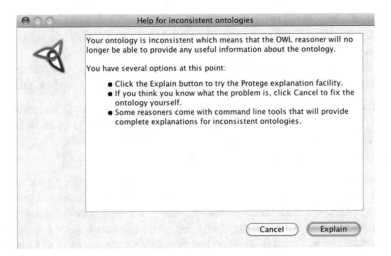

**Fig. 6.2** Protégé informs users about ontology inconsistency that prevents reasoning

Knowledge base inconsistency is considered a severe error, because reasoners cannot use inconsistent ontologies to infer useful information (see Fig. 6.2).

Common causes of inconsistency include the instantiation of an unsatisfiable class, the instantiation of disjoint classes, conflicting individual assertions in the ABox, conflicting axioms with nominals in the TBox, and instantiations that are not possible in the TBox [197].

If the consistency checking focuses on the ABox (with reference to the TBox) rather than the entire knowledge base, the inference is called *ABox consistency checking*.

Knowledge base consistency checking can be reduced via *internalization* to the satisfiability of concepts (see Definition 6.2).

**Definition 6.2 (Concept Satisfiability)** Given concept $C$ and knowledge base $\mathcal{K}$ as input, a decision procedure for concept satisfiability with reference to knowledge base $\mathcal{K}$ returns "$C$ is satisfiable with reference to $\mathcal{K}$" if there is an interpretation $\mathcal{I} = (\Delta^{\mathcal{I}}, \cdot^{\mathcal{I}})$ and an element $d \in \mathcal{I}$ such that $\mathcal{I} \vDash \mathcal{K}$ and $d \in C^{\mathcal{I}}$, otherwise it returns "$C$ is unsatisfiable with reference to $\mathcal{K}$."

The visualization of the hierarchy of the subclass-superclass relationships between concepts, called the *subsumption hierarchy*, provides the option to easily overview the knowledge domain model. Consequently, calculating the subsumption hierarchy (see Definition 6.3) is one of the most common reasoning tasks, and can be used to check whether the represented knowledge about subclasses is correct.

**Definition 6.3 (Concept Subsumption)** Concept $D$ subsumes concept $C$ with reference to knowledge base $\mathcal{K}$ iff $C^{\mathcal{I}} \subseteq D^{\mathcal{I}}$ for all interpretations $\mathcal{I}$ (that are models of $\mathcal{K}$).

In other words, given two complex concepts, $C$ and $D$, the decision procedure for concept subsumption determines whether the interpretation of $C$ is a subset of the interpretation of $D$, i.e., concept $D$ is more generic than concept $C$.

Another core reasoning task is *instance checking*, also known as *instantiation*, which checks whether an individual is an instance of a concept (see Definition 6.4).

**Definition 6.4 (Instance Checking)**  Given instance $a$, concept $C$, and knowledge base $\mathcal{K}$, a decision procedure for instance checking returns "$a$ is an instance of $C$ with reference to $\mathcal{K}$" if for each interpretation $\mathcal{I} = (\Delta^{\mathcal{I}}, {}^{\mathcal{I}})$ such that $\mathcal{I} \models \mathcal{K}$, $a^{\mathcal{I}} \in C^{\mathcal{I}}$, otherwise it returns "$a$ is not an instance of $C$ with reference to $\mathcal{K}$."

Reductions make it possible to implement only one reasoning procedure, in order to use any core reasoning service, which is exploited in most modern DL reasoners. Concept satisfiability, subsumption, and instance checking can be reduced to knowledge base satisfiability for all description logics that are closed under negation, i.e., for all description logics in which a complement of a concept is also a concept of the same description logic.

In description logics that support the negation of descriptions, the core reasoning tasks of subsumption, equivalence, and disjointness of concepts can be reduced to the concept (un)satisfiability problem (see Proposition 6.1).

**Proposition 6.1 (Reduction to Concept (Un)satisfiability)**  For concepts $C$ and $D$, $C \sqsubseteq_{\mathcal{T}} D \Leftrightarrow C \sqcap \neg D$ is unsatisfiable with reference to $\mathcal{T}$; $C \equiv D \Leftrightarrow (C \sqcap \neg D)$ and $(\neg C \sqcap D)$ are unsatisfiable; and $disjoint(C, D) \Leftrightarrow C \sqcap D$ is unsatisfiable.

For those expressive description logics that support all Boolean concept constructors, satisfiability and instantiation can be reduced to ABox (in)consistency with reference to a TBox (see Proposition 6.2), which can be executed in polynomial time.

**Proposition 6.2 (Reduction to ABox (In)consistency)**  In all description logics that support full negation, concept $C$ satisfiability can be reduced to consistency, i.e., $C$ is satisfiable with reference to $\mathcal{T} \Leftrightarrow$ ABox $\{C(a)\}$ is consistent with reference to $\mathcal{T}$, and instantiation can be reduced to inconsistency, i.e., $\mathcal{A} \models_{\mathcal{T}} C(a) \Leftrightarrow \mathcal{A} \cup \{\neg C (a)\}$ is inconsistent with reference to $\mathcal{T}$.

Some inference tasks can be reduced to subsumption reasoning (see Proposition 6.3).

**Proposition 6.3 (Reduction to Subsumption)**  For concepts $C$ and $D$, $C(a) \Leftrightarrow \{a\} \sqsubseteq_{\mathcal{T}} C$; $\nexists \mathcal{I} = (\Delta^{\mathcal{I}}, {}^{\mathcal{I}})$ and $\nexists d \in \mathcal{I}$ such that $\mathcal{I} \models \mathcal{K}$ and $d \in C^{\mathcal{I}} \Leftrightarrow C \sqsubseteq \bot$; $C \equiv_{\mathcal{T}} D \Leftrightarrow (C \sqsubseteq_{\mathcal{T}} D) \sqcap (D \sqsubseteq_{\mathcal{T}} C)$; $disjoint(C, D) \Leftrightarrow C \sqcap D \sqsubseteq \bot$.

A generalization of instance checking is *query answering*, which checks whether a given object is an instance of a specified concept in every model of a knowledge base (see Definition 6.5) [198].

**Definition 6.5 (Query Answering)**   Given a query $q$ with distinguished variables $\vec{x}$ and a knowledge base $\mathcal{K}$, return the set $ans(q, \mathcal{K})$ of tuples $\vec{c}$ of constants of $\mathcal{K}$ such that $\vec{c} \in q^{\mathcal{I}}$ in every model $\mathcal{I}$ of $\mathcal{K}$.

*Query containment* generalizes logical implication of inclusion assertions in description logics (see Definition 6.6).

**Definition 6.6 (Query Containment)**   Given two queries $q_1$ and $q_2$ and a KB $\mathcal{K}$, verify whether in every model $\mathcal{I}$ of $\mathcal{K}$ $q_1^{\mathcal{I}} \subseteq q_2^{\mathcal{I}}$.

A different kind of inference service is *retrieval inference*, which attempts to find individuals or roles that satisfy certain conditions (see Definition 6.7) [199].

**Definition 6.7 (Retrieval Inference)**   Given a knowledge base $\mathcal{K} = (\mathcal{T}, \mathcal{A}, \mathcal{R})$, a retrieval inference problem is to find all individuals of ABox $\mathcal{A}$ that are instances of a particular concept $C$.

Standard reasoning tasks, such as consistency and subsumption, are feasible in polynomial time even for all OWL 2 profiles (OWL 2 EL, OWL 2 QL, and OWL 2 RL) [200]. Several highly scalable profile-specific reasoners are also available, which are detailed later in the chapter.

## 6.3   Reasoning Rules

The level of semantic representation and mathematical formalization, together with the knowledge base size, presence or absence of instances, and the capabilities of the reasoner, determines what kind of reasoning is feasible. The reasoning tasks rely on different sets of reasoning rules, such as RDFS entailment, which gives semantics to the RDF and RDFS vocabularies; D-entailment, which provides semantics to datatypes for RDFS; and the OWL entailment, which adds semantics to the OWL vocabulary.[1]

### 6.3.1   RDFS Reasoning Rules

When using RDFS constructs, there are 14 entailment rules to infer new information (see Table 6.1) [201].

---

[1]Entailment is a logical consequence, where the conclusion is a consequence of, and logically follows from, the premise. On the Semantic Web, entailment is a fundamental concept to connect model-theoretic semantics to real-world objects.

**Table 6.1**  RDFS entailment rules

| If | Then |
|---|---|
| s p o (if o is a literal) | :n rdf:type rdfs:Literal |
| p rdfs:domain x and s p o | s rdf:type x |
| p rdfs:range x and s p o | o rdf:type x |
| s p o | s rdf:type rdfs:Resource |
| s p o | o rdf:type rdfs:Resource |
| p rdfs:subPropertyOf q and q rdfs:subPropertyOf r | p rdfs:subPropertyOf r |
| p rdf:type rdf:Property | p rdfs:subPropertyOf p |
| s p o and p rdfs:subPropertyOf q | s q o |
| s rdf:type rdfs:Class | s rdfs:subClassOf rdfs:Resource |
| s rdf:type x and x rdfs:subClassOf y | s rdf:type y |
| s rdf:type rdfs:Class | s rdfs:subClassOf s |
| x rdfs:subClassOf y and y rdfs:subClassof z | x rdfs:subClassOf z |
| p rdf:type rdfs:ContainerMembershipProperty | p rdfs:subPropertyOf rdfs:member |
| o rdf:type rdfs:Datatype | o rdfs:subClassOf rdfs:Literal |

## *6.3.2   Ter Horst Reasoning Rules*

The *pD\* semantics* proposed by Ter Horst, which inspired OWL 2 RL (beyond
*DLP*), takes a fragment of OWL Full into account, thereby achieving a low (NP or
P) complexity for the computation of the closure [202]. The *pD\** semantics are
expressed with a set of rules similar to, although more complex than, RDFS. The
*pD\** semantics cover datatypes and a subset of the OWL vocabulary, including
property-related terms (e.g., FunctionalProperty), comparisons (e.g.,
sameAs, differentFrom), and value restrictions (e.g., allValuesFrom).
The Ter Horst reasoning rules are summarized in Table 6.2.

**Table 6.2** The Ter Horst reasoning rules

| If | Then |
|---|---|
| `p rdf:type`<br>`owl:FunctionalProperty`<br>`u p v`<br>`u p w` | `v owl:sameAs w` |
| `p rdf:type`<br>`owl:InverseFunctionalProperty`<br>`v p u`<br>`w p u` | `v owl:sameAs w` |
| `p rdf:type`<br>`owl:SymmetricProperty`<br>`v p u` | `u p v` |
| `p rdf:type`<br>`owl:TransitiveProperty`<br>`u p w`<br>`w p v` | `u p v` |
| `u p v` | `u owl:sameAs u` |
| `u p v` | `v owl:sameAs v` |
| `v owl:sameAs w` | `w owl:sameAs v` |
| `v owl:sameAs w`<br>`w owl:sameAs u` | `v owl:sameAs u` |
| `p owl:inverseOf q`<br>`v p w` | `w q v` |
| `p owl:inverseOf q`<br>`v q w` | `w p v` |
| `v rdf:type owl:Class`<br>`v owl:sameAs w` | `v rdfs:subClassOf w` |
| `p rdf:type owl:Property`<br>`p owl:sameAs q` | `p rdfs:subPropertyOf q` |
| `u p v`<br>`u owl:sameAs x`<br>`v owl:sameAs y` | `x p y` |
| `v owl:equivalentClass w` | `v rdfs:subClassOf w` |
| `v owl:equivalentClass w` | `w rdfs:subClassOf v` |
| `v rdfs:subClassOf w`<br>`w rdfs:subClassOf v` | `v rdfs:equivalentClass w` |
| `v owl:equivalentProperty w` | `v rdfs:subPropertyOf w` |
| `v owl:equivalentProperty w` | `w rdfs:subPropertyOf v` |
| `v rdfs:subPropertyOf w`<br>`w rdfs:subPropertyOf v` | `v rdfs:equivalentProperty w` |

(continued)

**Table 6.2** (continued)

| | |
|---|---|
| `v owl:hasValue w`<br>`v owl:onProperty p`<br>`u p v` | `u rdf:type v` |
| `v owl:hasValue w`<br>`v owl:onProperty p`<br>`u rdf:type v` | `u p v` |
| `v owl:someValuesFrom w`<br>`v owl:onProperty p`<br>`u p x`<br>`x rdf:type w` | `u rdf:type v` |
| `v owl:allValuesFrom w`<br>`v owl:onProperty p`<br>`u rdf:type v`<br>`u p x` | `x rdf:type w` |

## *6.3.3   OWL 2 Reasoning Rules*

*OWL 2 RDF-based semantics* are the RDF-compatible model-theoretic semantics of OWL 2, which give formal meaning to RDF graphs. The OWL 2 RDF-based semantics are defined as semantic extensions of D-entailment (i.e., RDFS with datatype support). Consequently, the meaning of an RDF graph given by the OWL 2 RDF-based semantics consists of the meaning provided by the semantics of D-entailment and the additional meaning specified for all the language constructors of OWL 2, such as Boolean connectives, subproperty chains, and qualified cardinality restrictions [203].

The *OWL 2 RL/RDF rules* partially axiomatize the OWL 2 RDF-based semantics in the form of first-order implications, and are defined as universally quantified first-order implications over the ternary predicate `T` of the form `T(s, p, o)`, which represents a generalization of RDF triples with blank nodes and literals in arbitrary positions [204]. In the implications, each variable starts with a question mark. The rules without an "if" part are considered always applicable.

Table 6.3 defines the reflexivity, symmetry, and transitivity of the equality relation `owl:sameAs`, and axiomatizes the standard replacement properties of equality.

**Table 6.3** Reasoning rules for equality

| If | Then |
|---|---|
| `T(?s, ?p, ?o)` | `T(?s, owl:sameAs, ?s)`<br>`T(?p, owl:sameAs, ?p)`<br>`T(?o, owl:sameAs, ?o)` |
| `T(?x, owl:sameAs, ?y)` | `T(?y, owl:sameAs, ?x)` |
| `T(?x, owl:sameAs, ?y)`<br>`T(?y, owl:sameAs, ?z)` | `T(?x, owl:sameAs, ?z)` |
| `T(?s, owl:sameAs, ?s')`<br>`T(?s, ?p, ?o)` | `T(?s', ?p, ?o)` |
| `T(?p, owl:sameAs, ?p')`<br>`T(?s, ?p, ?o)` | `T(?s, ?p', ?o)` |
| `T(?o, owl:sameAs, ?o')`<br>`T(?s, ?p, ?o)` | `T(?s, ?p, ?o')` |
| `T(?x, owl:sameAs, ?y)`<br>`T(?x, owl:differentFrom, ?y)` | `false` |
| `T(?x, rdf:type, owl:AllDifferent)`<br>`T(?x, owl:members, ?y)`<br>`LIST[?y, ?z1, …, ?zn]`<br>`T(?zi, owl:sameAs, ?zj)` | `false` for each $1 \leqslant i < j \leqslant n$ |
| `T(?x, rdf:type, owl:AllDifferent)`<br>`T(?x, owl:distinctMembers, ?y)`<br>`LIST[?y, ?z1, …, ?zn]`<br>`T(?zi, owl:sameAs, ?zj)` | `false` for each $1 \leqslant i < j \leqslant n$ |

The semantic conditions on classes are listed in Table 6.4.

**Table 6.4** Reasoning rules for classes

| If | Then |
|---|---|
|  | T(owl:Thing, rdf:type, owl:Class) |
|  | T(owl:Nothing, rdf:type, owl:Class) |
| T(?x, rdf:type, owl:Nothing) | false |
| T(?c, owl:intersectionOf, ?x)<br>LIST[?x, ?c1, …, ?cn]<br>T(?y, rdf:type, ?c1)<br>T(?y, rdf:type, ?c2)<br>…<br>T(?y, rdf:type, ?cn) | T(?y, rdf:type, ?c) |
| T(?c, owl:intersectionOf, ?x)<br>LIST[?x, ?c1, …, ?cn]<br>T(?y, rdf:type, ?c) | T(?y, rdf:type, ?c1)<br>T(?y, rdf:type, ?c2)<br>…<br>T(?y, rdf:type, ?cn) |
| T(?c, owl:unionOf, ?x)<br>LIST[?x, ?c1, …, ?cn]<br>T(?y, rdf:type, ?ci) | T(?y, rdf:type, ?c)<br>for each $1 \leqslant i \leqslant n$ |
| T(?c1, owl:complementOf, ?c2)<br>T(?x, rdf:type, ?c1)<br>T(?x, rdf:type, ?c2) | false |
| T(?x, owl:someValuesFrom, ?y)<br>T(?x, owl:onProperty, ?p)<br>T(?u, ?p, ?v)<br>T(?v, rdf:type, ?y) | T(?u, rdf:type, ?x) |
| T(?x, owl:someValuesFrom, owl:Thing)<br>T(?x, owl:onProperty, ?p) | T(?u, rdf:type, ?x) |

(continued)

**Table 6.4** (continued)

| T(?u, ?p, ?v) | |
|---|---|
| T(?x, owl:allValuesFrom, ?y)<br>T(?x, owl:onProperty, ?p)<br>T(?u, rdf:type, ?x)<br>T(?u, ?p, ?v) | T(?v, rdf:type, ?y) |
| T(?x, owl:hasValue, ?y)<br>T(?x, owl:onProperty, ?p)<br>T(?u, rdf:type, ?x) | T(?u, ?p, ?y) |
| T(?x, owl:hasValue, ?y)<br>T(?x, owl:onProperty, ?p)<br>T(?u, ?p, ?y) | T(?u, rdf:type, ?x) |
| T(?x, owl:maxCardinality,<br>"0"^^xsd:nonNegativeInteger)<br>T(?x, owl:onProperty, ?p)<br>T(?u, rdf:type, ?x)<br>T(?u, ?p, ?y) | false |
| T(?x, owl:maxCardinality,<br>"1"^^xsd:nonNegativeInteger)<br>T(?x, owl:onProperty, ?p)<br>T(?u, rdf:type, ?x)<br>T(?u, ?p, ?y1)<br>T(?u, ?p, ?y2) | T(?y1, owl:sameAs, ?y2) |
| T(?x,<br>owl:maxQualifiedCardinality,<br>"0"^^xsd:nonNegativeInteger)<br>T(?x, owl:onProperty, ?p)<br>T(?x, owl:onClass, ?c)<br>T(?u, rdf:type, ?x)<br>T(?u, ?p, ?y)<br>T(?y, rdf:type, ?c) | false |
| T(?x,<br>owl:maxQualifiedCardinality,<br>"0"^^xsd:nonNegativeInteger)<br>T(?x, owl:onProperty, ?p)<br>T(?x, owl:onClass,<br>owl:Thing)<br>T(?u, rdf:type, ?x)<br>T(?u, ?p, ?y) | false |

(continued)

**Table 6.4** (continued)

| | |
|---|---|
| T(?x, owl:maxQualifiedCardinality, "1"^^xsd:nonNegativeInteger) T(?x, owl:onProperty, ?p) T(?x, owl:onClass, ?c) T(?u, rdf:type, ?x) T(?u, ?p, ?y1) T(?y1, rdf:type, ?c) T(?u, ?p, ?y2) T(?y2, rdf:type, ?c) | T(?y1, owl:sameAs, ?y2) |
| T(?x, owl:maxQualifiedCardinality, "1"^^xsd:nonNegativeInteger) T(?x, owl:onProperty, ?p) T(?x, owl:onClass, owl:Thing) T(?u, rdf:type, ?x) T(?u, ?p, ?y1) T(?u, ?p, ?y2) | T(?y1, owl:sameAs, ?y2) |
| T(?c, owl:oneOf, ?x) LIST[?x, ?y1, …, ?yn] | T(?y1, rdf:type, ?c) … T(?yn, rdf:type, ?c) |

Table 6.5 defines the semantic conditions on class axioms.

**Table 6.5**  Reasoning rules for class axioms

| If | Then |
|---|---|
| `T(?c1, rdfs:subClassOf, ?c2)`<br>`T(?x, rdf:type, ?c1)` | `T(?x, rdf:type, ?c2)` |
| `T(?c1, owl:equivalentClass, ?c2)`<br>`T(?x, rdf:type, ?c1)` | `T(?x, rdf:type, ?c2)` |
| `T(?c1, owl:equivalentClass, ?c2)`<br>`T(?x, rdf:type, ?c2)` | `T(?x, rdf:type, ?c1)` |
| `T(?c1, owl:disjointWith, ?c2)`<br>`T(?x, rdf:type, ?c1)`<br>`T(?x, rdf:type, ?c2)` | `false` |
| `T(?x, rdf:type, owl:AllDisjointClasses)`<br>`T(?x, owl:members, ?y)`<br>`LIST[?y, ?c1, …, ?cn]`<br>`T(?z, rdf:type, ?ci)`<br>`T(?z, rdf:type, ?cj)` | `false` for each $1 \leqslant i < j \leqslant n$ |

The semantic conditions on axioms about properties are summarized in Table 6.6.

**Table 6.6** Reasoning rules for axioms about properties

| If | Then |
|---|---|
| | T(ap, rdf:type, owl:AnnotationProperty) |
| T(?p, rdfs:domain, ?c)<br>T(?x, ?p, ?y) | T(?x, rdf:type, ?c) |
| T(?p, rdfs:range, ?c)<br>T(?x, ?p, ?y) | T(?y, rdf:type, ?c) |
| T(?p, rdf:type,<br>owl:FunctionalProperty)<br>T(?x, ?p, ?y1)<br>T(?x, ?p, ?y2) | T(?y1, owl:sameAs, ?y2) |
| T(?p, rdf:type,<br>owl:InverseFunctionalProperty)<br>T(?x1, ?p, ?y)<br>T(?x2, ?p, ?y) | T(?x1, owl:sameAs, ?x2) |
| T(?p, rdf:type,<br>owl:IrreflexiveProperty)<br>T(?x, ?p, ?x) | false |
| T(?p, rdf:type,<br>owl:SymmetricProperty)<br>T(?x, ?p, ?y) | T(?y, ?p, ?x) |
| T(?p, rdf:type,<br>owl:AsymmetricProperty)<br>T(?x, ?p, ?y)<br>T(?y, ?p, ?x) | false |
| T(?p, rdf:type,<br>owl:TransitiveProperty)<br>T(?x, ?p, ?y)<br>T(?y, ?p, ?z) | T(?x, ?p, ?z) |
| T(?p1, rdfs:subPropertyOf,<br>?p2)<br>T(?x, ?p1, ?y) | T(?x, ?p2, ?y) |
| T(?p, owl:propertyChainAxiom,<br>?x)<br>LIST[?x, ?p1, …, ?pn]<br>T(?u1, ?p1, ?u2)<br>T(?u2, ?p2, ?u3)<br>…<br>T(?un, ?pn, ?un+1) | T(?u1, ?p, ?un+1) |

(continued)

**Table 6.6** (continued)

| | |
|---|---|
| T(?p1, owl:equivalentProperty, ?p2)<br>T(?x, ?p1, ?y) | T(?x, ?p2, ?y) |
| T(?p1, owl:equivalentProperty, ?p2)<br>T(?x, ?p2, ?y) | T(?x, ?p1, ?y) |
| T(?p1, owl:propertyDisjointWith, ?p2)<br>T(?x, ?p1, ?y)<br>T(?x, ?p2, ?y) | false |
| T(?x, rdf:type, owl:AllDisjointProperties)<br>T(?x, owl:members, ?y)<br>LIST[?y, ?p1, …, ?pn]<br>T(?u, ?pi, ?v)<br>T(?u, ?pj, ?v)<br><br>for each $1 \leqslant i < j \leqslant n$ | false |
| T(?p1, owl:inverseOf, ?p2)<br>T(?x, ?p1, ?y) | T(?y, ?p2, ?x) |
| T(?p1, owl:inverseOf, ?p2)<br>T(?x, ?p2, ?y) | T(?y, ?p1, ?x) |
| T(?c, owl:hasKey, ?u)<br>LIST[?u, ?p1, …, ?pn]<br>T(?x, rdf:type, ?c)<br>T(?x, ?p1, ?z1)<br>…<br>T(?x, ?pn, ?zn)<br>T(?y, rdf:type, ?c)<br>T(?y, ?p1, ?z1)<br>…<br>T(?y, ?pn, ?zn) | T(?x, owl:sameAs, ?y) |
| T(?x, owl:sourceIndividual, ?i1)<br>T(?x, owl:assertionProperty, ?p)<br>T(?x, owl:targetIndividual, ?i2)<br>T(?i1, ?p, ?i2) | false |
| T(?x, owl:sourceIndividual, ?i)<br>T(?x, owl:assertionProperty, ?p)<br>T(?x, owl:targetValue, ?lt)<br>T(?i, ?p, ?lt) | false |

Table 6.7 specifies the semantic restrictions of schema vocabulary.

**Table 6.7**   Reasoning rules for schema vocabulary

| If | Then |
|---|---|
| `T(?c, rdf:type, owl:Class)` | `T(?c, rdfs:subClassOf, ?c)`<br>`T(?c, owl:equivalentClass, ?c)`<br>`T(?c, rdfs:subClassOf,`<br>`owl:Thing)`<br>`T(owl:Nothing, rdfs:subClassOf,`<br>`?c)` |
| `T(?c1, rdfs:subClassOf,`<br>`?c2)`<br>`T(?c2, rdfs:subClassOf,`<br>`?c3)` | `T(?c1, rdfs:subClassOf, ?c3)` |
| `T(?c1,`<br>`owl:equivalentClass, ?c2)` | `T(?c1, rdfs:subClassOf, ?c2)`<br>`T(?c2, rdfs:subClassOf, ?c1)` |
| `T(?c1, rdfs:subClassOf,`<br>`?c2)`<br>`T(?c2, rdfs:subClassOf,`<br>`?c1)` | `T(?c1, owl:equivalentClass, ?c2)` |
| `T(?p, rdf:type,`<br>`owl:ObjectProperty)` | `T(?p, rdfs:subPropertyOf, ?p)`<br>`T(?p, owl:equivalentProperty,`<br>`?p)` |
| `T(?p, rdf:type,`<br>`owl:DatatypeProperty)` | `T(?p, rdfs:subPropertyOf, ?p)`<br>`T(?p, owl:equivalentProperty,`<br>`?p)` |
| `T(?p1, rdfs:subPropertyOf,`<br>`?p2)`<br>`T(?p2, rdfs:subPropertyOf,`<br>`?p3)` | `T(?p1, rdfs:subPropertyOf, ?p3)` |
| `T(?p1,`<br>`owl:equivalentProperty,`<br>`?p2)` | `T(?p1, rdfs:subPropertyOf, ?p2)`<br>`T(?p2, rdfs:subPropertyOf, ?p1)` |
| `T(?p1, rdfs:subPropertyOf,`<br>`?p2)`<br>`T(?p2, rdfs:subPropertyOf,`<br>`?p1)` | `T(?p1, owl:equivalentProperty,`<br>`?p2)` |
| `T(?p, rdfs:domain, ?c1)`<br>`T(?c1, rdfs:subClassOf,`<br>`?c2)` | `T(?p, rdfs:domain, ?c2)` |
| `T(?p2, rdfs:domain, ?c)`<br>`T(?p1, rdfs:subPropertyOf,`<br>`?p2)` | `T(?p1, rdfs:domain, ?c)` |

(continued)

**Table 6.7** (continued)

| | |
|---|---|
| T(?p, rdfs:range, ?c1)<br>T(?c1, rdfs:subClassOf,<br>?c2) | T(?p, rdfs:range, ?c2) |
| T(?p2, rdfs:range, ?c)<br>T(?p1, rdfs:subPropertyOf,<br>?p2) | T(?p1, rdfs:range, ?c) |
| T(?c1, owl:hasValue, ?i)<br>T(?c1, owl:onProperty,<br>?p1)<br>T(?c2, owl:hasValue, ?i)<br>T(?c2, owl:onProperty,<br>?p2)<br>T(?p1, rdfs:subPropertyOf,<br>?p2) | T(?c1, rdfs:subClassOf, ?c2) |
| T(?c1, owl:someValuesFrom,<br>?y1)<br>T(?c1, owl:onProperty, ?p)<br>T(?c2, owl:someValuesFrom,<br>?y2)<br>T(?c2, owl:onProperty, ?p)<br>T(?y1, rdfs:subClassOf,<br>?y2) | T(?c1, rdfs:subClassOf, ?c2) |
| T(?c1, owl:someValuesFrom,<br>?y)<br>T(?c1, owl:onProperty,<br>?p1)<br>T(?c2, owl:someValuesFrom,<br>?y)<br>T(?c2, owl:onProperty,<br>?p2)<br>T(?p1, rdfs:subPropertyOf,<br>?p2) | T(?c1, rdfs:subClassOf, ?c2) |
| T(?c1, owl:allValuesFrom,<br>?y1)<br>T(?c1, owl:onProperty, ?p)<br>T(?c2, owl:allValuesFrom,<br>?y2)<br>T(?c2, owl:onProperty, ?p)<br>T(?y1, rdfs:subClassOf,<br>?y2) | T(?c1, rdfs:subClassOf, ?c2) |

(continued)

**Table 6.7** (continued)

| | |
|---|---|
| T(?c1, owl:allValuesFrom, ?y)<br>T(?c1, owl:onProperty, ?p1)<br>T(?c2, owl:allValuesFrom, ?y)<br>T(?c2, owl:onProperty, ?p2)<br>T(?p1, rdfs:subPropertyOf, ?p2) | T(?c2, rdfs:subClassOf, ?c1) |
| T(?c, owl:intersectionOf, ?x)<br>LIST[?x, ?c1, …, ?cn] | T(?c, rdfs:subClassOf, ?c1)<br>T(?c, rdfs:subClassOf, ?c2)<br>…<br>T(?c, rdfs:subClassOf, ?cn) |
| T(?c, owl:unionOf, ?x)<br>LIST[?x, ?c1, …, ?cn] | T(?c1, rdfs:subClassOf, ?c)<br>T(?c2, rdfs:subClassOf, ?c)<br>…<br>T(?cn, rdfs:subClassOf, ?c) |

The semantics of datatypes can be briefly summarized as follows. T(dt, rdf:type, rdfs:Datatype) holds for each datatype dt supported in OWL 2 RL. T(lt, rdf:type, dt) holds for each literal lt and each datatype dt supported in OWL 2 RL such that the data value of lt is contained in the value space of dt. T(lt1, owl:sameAs, lt2) holds for all literals lt1 and lt2 with the same data value. T(lt1, owl:differentFrom, lt2) holds for all literals lt1 and lt2 with different data values. If T(lt, rdf:type, dt) holds, then false applies to each literal lt and each datatype dt supported in OWL 2 RL such that the data value of lt is not contained in the value space of dt.

## 6.4  DL Reasoning Algorithms

Description logics employ different types of logical inference, such as deduction and abduction. *Deductive reasoning* (deduction) is reasoning from one or more premises (statements) to reach a logically certain conclusion. If all premises are true and the terms are clear, the conclusion of deduction is necessarily true. Deduction is the default reasoning type of description logics.

Inference services different from the standard (deductive) reasoning tasks have also been introduced, such as *abductive reasoning* (see Definition 6.8).

**Definition 6.8 (Abductive Reasoning)** An inference task is called abduction if it attempts to construct a set of minimal explanations $E$ for a given set of assertions $\Gamma$ such that $E$ is consistent with reference to knowledge base $\mathcal{K} = (\mathcal{T}, \mathcal{A}, \mathcal{R})$ and satisfies $\mathcal{T} \cup \mathcal{A} \cup \mathcal{R} \cup E \models \top$ and if $E'$ is an ABox satisfying $\mathcal{T} \cup \mathcal{A} \cup \mathcal{R} \cup E' \models \Gamma$, then $E' \models E$, i.e., $E$ is least specific.

Abductive reasoning (abduction) attempts to find the simplest and most likely explanation while progressing from an observation to a theory. Abduction can be used for multimedia interpretation by generating potential explanations based on background knowledge formally represented in knowledge bases and context derived from ABox axioms, and choosing the most likely explanation based on the maxima of preference scores. However, abduction is inherently uncertain, so the premises do not guarantee the conclusion.

*Inductive reasoning* (induction) is reasoning in which the premises are viewed as supplying strong evidence for the truth of the conclusion. In contrast to the conclusion of deduction, which is certain, the truth of the conclusion of inductive reasoning is only probable, depending on the evidence given.

The following sections describe the most common decision procedures used in DL reasoning, such as automata, tableau, and resolution calculi.

## 6.4.1 Tableau-Based Consistency Checking

Most OWL reasoners, such as FaCT++, Pellet, and Racer, are based on highly optimized *tableau* algorithms, which attempt to construct an abstraction of a tree-like model $\mathcal{I}$ (a *completion graph* chosen in a particular iteration) that satisfies all axioms of a knowledge base, thereby proving (un)satisfiability. A tableau contains all the relevant information from a model, with the possible exception of transitive roles ($t$) and their superroles ($s$). Tableau algorithms have been proposed in the literature for the most common description logics, including very expressive ones. Consequently, the formal description of a tableau algorithm depends on the description logic for which it was designed. Generally speaking, tableau algorithms work on concepts in *negation normal form* (NNF) (see Definition 6.9), in which the only two allowed Boolean operators are conjunction and disjunction, by using de Morgan's laws and exploiting equivalences such as $\neg\neg C \equiv C$, $\neg(C \sqcap D) \equiv (\neg C) \sqcup (\neg D)$, $\neg(C \sqcup D) \equiv (\neg C) \sqcap (\neg D)$, $\neg(\exists R.C) \equiv (\forall R.\neg C)$, $\neg(\forall R.C) \equiv (\exists R.\neg C)$, $\neg(\leqslant nR.C) \equiv (\geqslant (n+1)R.C)$, $\neg(\geqslant (n+1)R.C) \equiv (\leqslant nR.C)$, and $\neg(\geqslant 0R.C) \equiv \bot$.

**Definition 6.9 (Negation Normal Form of a Concept)** A concept $C$ is in negation normal form if the negation operator $\neg$ only appears in front of concept names.

A concept $C$ is broken down syntactically by inferring constraints on elements of $\mathcal{I}$. The decomposition step utilizes the following *derivation rules* [205]:

- $\sqcup$-rule: given $(C_1 \sqcup C_2)(s)$, derive either $C_1(s)$ or $C_2(s)$
- $\sqcap$-rule: given $(C_1 \sqcap C_2)(s)$, derive $C_1(s)$ and $C_2(s)$

- $\exists$-rule: given $(\exists R.C(s))$, derive $R(s,t)$ and $C(t)$ for $t$ a fresh individual
- $\forall$-rule: given $(\forall R.C)(s)$ and $R(s,t)$, derive $C(t)$
- $\sqsubseteq$-rule: given a general concept inclusion of the form $C \sqsubseteq D$ and an individual $s$, derive $(\neg C \sqcup D)(s)$

The set of elements is created based on the ABox axioms, and used to retrieve concept memberships and role assertions. The constructed intermediate model often does not satisfy every TBox and RBox axiom, so the model has to be updated accordingly in each step. As a result, new concept memberships and role relationships might be created. When case distinction occurs, the algorithm might have to backtrack. If a state is reached when all axioms are satisfied, the ontology is considered satisfiable. OWL 2 reasoners, such as HermiT, usually use a tableau refinement based on the *hypertableau* and *hyperresolution* calculi to reduce the nondeterminism caused by general inclusion axioms [206].

To demonstrate integrity checking using the tableau algorithm, assume the following axioms:

| | |
|---|---|
| depicts $\sqsubseteq$ features | (A1) |
| Gunfight $\sqsubseteq$ ¬Chase | (A2) |
| Action $\sqsubseteq$ Chase $\sqcup$ Combat | (A3) |
| ViolentActionScene $\sqsubseteq$ $\exists$features.Action $\sqcap$ $\forall$features.Gunfight | (A4) |
| ViolentActionScene(SCENE1) | (A5) |

Based on the only ABox axiom (A5), tableau-based reasoners would consider SCENE1 as a violent action scene (SCENE1 is an individual of the Violent-ActionScene concept); however, this temporary representation would not satisfy A4. Next, reasoners would introduce a new individual. The connection between the individual (SCENE1) and the new individual is defined with the depicts predicate. As a result, the definition of violent action scenes (A4) is now satisfied; however, other TBox axioms are invalidated (A1 and A3). To address this issue, reasoners would introduce a connection from the individual (SCENE1) to the new individual. Finally, a case distinction is needed, because an action can be either a chase or a combat. In the first case, A4 is violated because of the second part of its consequence. To address this issue, the new individual has to be marked with Gunfight, which then invalidates A2, meaning that the new individual must be ¬Chase. Because the new individual cannot be marked with both Chase and ¬Chase, the algorithm needs to backtrack. In the second case, the new individual is marked as Combat, which violates A4, therefore the new individual must be marked with Gunfight, which invalidates A2. Consequently, the new individual is marked as ¬Chase, which leads to a knowledge representation model that satisfies all axioms, upon which reasoners can conclude that the knowledge base is satisfiable.

### 6.4.2   Automata

Automaton-based methods have been successfully employed in description logics to prove tight complexity bounds for reasoning [207]. By creating an automaton that characterizes the tree models of a knowledge base, the problem of knowledge base satisfiability can be reduced to checking the nonemptiness of the tree language represented by the corresponding automaton [208].

### 6.4.3   Resolution

One of the most common calculi for theorem proving in first-order logic is *resolution*. A resolution that involves more than two input clauses is called *hyperresolution*. Resolution proves a theorem by negating the statement to prove and adding this negated goal to the sets of axioms that are known to be true. Finally, it employs hyperresolution (see Expression 6.1), positive factoring (see Expression 6.2), and inference rules to show that this leads to contradiction [209].

$$\frac{C_1 \vee A_1 ... C_n \vee A_n \neg A_{n+1} \vee ... \vee \neg A_{2n} \vee D}{(C_1 \vee ... \vee C_n \vee D)\sigma} \tag{6.1}$$

where $\sigma$ is the most general unifier such that $A_i\sigma = A_{n+i}\sigma$ for every $i$, $1 \leqslant i \leqslant n$, $C_i \vee A_i$ and $D$ are positive clauses, for every $1 \leqslant i \leqslant n$. The rightmost premise in the rule is the negative premise and all other premises are positive premises.

$$\frac{C \vee A_1 \vee A_2}{(C \vee A_1)\sigma} \tag{6.2}$$

where $\sigma$ is the most general unifier of $A_1$ and $A_2$.

Resolution is sound and complete. Because the resolution rule is a deduction rule, resolution can be considered as a variant of consequence-based reasoning, although resolution is performed on the first-order logic underpinning or translation of a knowledge base, not the DL knowledge base itself.

## 6.5   Reasoning Complexity

Supporting a large set of mathematical constructors leads to high expressivity, but at the same time increases computational complexity. By analyzing the computational complexity of description logics, their worst-case behavior can be characterized. For complexity analysis, the different types of computational complexities have to be distinguished. *Taxonomic complexity* is measured with respect to the total size of axioms without assertions in an ontology. *Data complexity* is measured with respect to the total size of assertions in an ontology, i.e., the ABox is the only input considered, and the size of the TBox, the role hierarchy, and the query, which is often significantly smaller than the size of the ABox, is fixed. Data complexity of a reasoning problem is a useful performance estimate, because it indicates the behavior of the reasoning algorithm in relation to the number of individual assertions in the ABox. *Query complexity* is measured with respect to the total size of the conjunctive query (defined only for conjunctive query answering). *Combined complexity* is measured with respect to the total size of axioms and assertions in an ontology, and also the conjunctive query (in the case of conjunctive query answering), or the expression(s) being checked (in the case of class expression satisfiability, class expression subsumption, or instance checking). Table 6.8 summarizes these complexities for some important description logics.

**Table 6.8** Reasoning complexity of common description logics

|  | Concept satisfiability checking complexity | ABox consistency checking complexity | Data complexity | Combined complexity |
|---|---|---|---|---|
| $\mathcal{EL}^{++}$ | PTIME | PTIME | PTIME | PTIME |
| $\mathcal{ALC}$ | PSPACE-complete (empty or acyclic TBox); EXPTIME-complete (general TBox) | PSPACE-complete (empty or acyclic TBox); EXPTIME-complete (general TBox) | co-NP-complete | EXPTIME-complete |
| $\mathcal{SHOIN}$ | NEXPTIME-complete | NEXPTIME-complete | co-NP-hard | NEXPTIME-hard |
| $\mathcal{SROIQ}$ | NEXPTIME-hard | NEXPTIME-hard | NP-hard | N2EXPTIME-complete |

The PSPACE complexity class is the set of decision problems that can be solved by a deterministic Turing machine using a polynomial amount of memory space. PTIME is a set of decision problems that can be solved by a deterministic Turing machine using a polynomial amount of computation time (polynomial time). EXPTIME is a set of decision problems that have exponential runtime, i.e., that are solvable by a deterministic Turing machine in $O^{(2p(n))}$ time, where $p(n)$ is a polynomial function of $n$. NEXPTIME is the set of decision problems that can be solved by a nondeterministic Turing machine using time $2^{n^{O(1)}}$, and N2EXPTIME is the set of decision problems that can be solved by a deterministic Turing machine with time bound $O(2^{2p(n)})$, as per computational complexity theory annotations. X-hard means at least as hard as noted (could be harder), in X means no harder than noted (could be easier); and X-complete means both hard and in, i.e., exactly as hard as noted. A decision problem is a member of co-NP if and only if its complement is in the complexity class NP

As Table 6.8 suggests, some description logic constructors increase computational complexity in parallel with improved expressivity; however, there are some DL constructors that do not affect reasoning. For example, adding arbitrary Boolean role constructors on simple roles to any description logic beyond $\mathcal{ALCIFV}$ does not increase data or combined complexity. Conjunctions on simple roles added to $\mathcal{ELV}^{++}$ do not increase data complexity or combined complexity. Adding concept products does not affect complexity either. In contrast to data complexity, which is co-NP-complete for most description logics from $DL\text{-}Lite_{core}$ to $\mathcal{ALCHOQ}$ [210], combined complexity varies greatly. Note that the reasoning complexity of concept satisfiability checking depends on whether the TBox is an empty, cyclic, acyclic, or general TBox. Reasoning complexity is also determined by whether there are closed predicates (predicates explicitly assumed complete) in a knowledge base, which can cause the loss of tractability.

## 6.6   DL-Based Reasoners

Reasoners derive new facts from existing ontologies and check the integrity of ontologies. The various software tools are different in terms of reasoning characteristics, practical usability, and performance, owing to the different algorithms implemented for description logic reasoning. Not all reasoners can evaluate all possible inferences, so their soundness and completeness vary. Not all reasoners can handle rules either. A common feature of reasoners is ABox reasoning, the reasoning of individuals that covers instance checking, conjunctive query answering, and consistency checking. Advanced reasoners support the OWL API, a standard interface for application development with OWL reasoning. Another feature of advanced reasoners is OWLLink support, which leverage an implementation-neutral protocol to interact with OWL 2 reasoners.

Some reasoners, such as the *Euler YAP Engine* (EYE),[2] only partially cover RDFS and OWL reasoning rules, while others provide a complete or almost complete coverage for OWL Lite (*Jena*),[3] OWL DL (*SHER* [211], *KAON 2*),[4] OWL 2 DL Lite (*Owlgres*),[5] OWL 2 DL (*HermiT, Pellet, FaCT++, Racer*), OWL 2 EL (*CEL*,[6] *Snorocket*),[7] OWL 2 RL (*FuXi*),[8] and OWL 2 QL (*Mastro*).[9] Some reasoners support a combination of OWL or OWL 2 profiles, such as *ELLY*[10] (OWL

---

[2]http://eulersharp.sourceforge.net

[3]http://jena.apache.org

[4]http://kaon2.semanticweb.org

[5]http://semanticweb.org/wiki/Owlgres.html

[6]http://lat.inf.tu-dresden.de/systems/cel/

[7]https://github.com/aehrc/snorocket

[8]https://code.google.com/archive/p/fuxi/

[9]http://www.dis.uniroma1.it/~mastro/

[10]http://elly.sourceforge.net/

2 EL and OWL 2 RL) and *OWLIM*[11] (OWL 2 QL and OWL 2 RL). *TrOWL*[12] utilizes a semantic approximation to transform OWL 2 DL ontologies into OWL 2 QL for conjunctive query answering, and a syntactic approximation from OWL 2 DL to OWL 2 EL for TBox and ABox reasoning. The *fuzzyDL*[13] reasoner was designed for fuzzy $\mathcal{SHIF}$ ontologies, but can also be used for other standard description logic reasoning tasks. The following sections focus on the most common description logic reasoners.

## 6.6.1   HermiT

*HermiT*[14] is one of the most popular and efficient OWL 2 reasoners that can perform all core reasoning tasks. HermiT implements a proprietary algorithm, called the *hypertableau calculus*, to check the consistency of OWL ontologies and identify subsumption relationships between classes. HermiT can be used as a Protégé plugin, through the command line, or in Java applications. The latest Protégé versions come with a preinstalled HermiT plugin. From the command line, one can perform classification, querying, and other common reasoning tasks. HermiT supports the OWLReasoner interface from the OWL API, providing programmatic access to OWL API objects, such as ontologies and class expressions.

HermiT employs several optimizations to reduce tableau complexity [212]. For example, it handles *or-branching* (disjunctions through reasoning by case), which is mainly caused by the $\sqsubseteq$-rule, because it adds a disjunction for each TBox axiom to each individual in an ABox. Because the processing of TBox axioms in tableau algorithms can be unnecessarily nondeterministic, HermiT implements absorption optimizations, including basic absorption, role absorption, and binary absorption. The *basic absorption* algorithm attempts to rewrite TBox axioms into the form $A \sqsubseteq C$, where $A$ is an atomic concept. Then, instead of deriving $\neg A \sqcup C$ for each individual in an ABox, $C(s)$ is derived only if the ABox contains $A(s)$. *Role absorption* rewrites axioms into the form $\exists R.\top \sqsubseteq C$, so that $C(s)$ is derived only if an ABox contains $R(s, t)$. *Binary absorption* rewrites general concept inclusions into the form $A_1 \sqcap A_2 \sqsubseteq C$, where $A_1, A_2$ are new internal primitive concepts introduced by absorption, so that $C(s)$ is derived only if an ABox contains both $A_1(s)$ and $A_2(s)$. Note that the combination of transformation and absorption techniques is usually based on heuristics and does not always eliminate nondeterminism completely. In HermiT, absorption optimizations are generalized by rewriting description logic axioms into a form that allows the simultaneous use of standard

---

[11]http://ontotext.com/owlim/

[12]http://trowl.org

[13]http://www.umbertostraccia.it/cs/software/fuzzyDL/fuzzyDL.html

[14]http://hermit-reasoner.com

absorption, role absorption, and binary absorption, and allows additional absorption types not available in standard tableau calculi.

HermiT also handles *and-branching* (the introduction of new individuals by the ∃-rule), which is another major source of inefficiency in tableau algorithms. To ensure termination, tableau algorithms employ *blocking*—if the two individuals *a* and *b* are identical, then *a* directly blocks *b* and no further expansion rules need to be applied to *a*. In contrast to standard tableau algorithms, which only allow individuals to be blocked by their ancestors (ancestor blocking), HermiT implements an improved approach called *anywhere blocking*. This can exponentially reduce the size of generated models, and substantially improve real-world performance on large and complex ontologies.

### 6.6.2  Pellet

Clark & Parsia's *Pellet*[15] is an OWL 2 DL reasoner, which can be used in Protégé, Jena, TopBraid Composer, or in Java programs through the OWL API interface. It is based on the tableau algorithm to break down complex statements into smaller and simpler pieces to detect contradictions and supports expressive description logics. Pellet supports different incremental reasoning, including incremental consistency checking and incremental classification, where updates (additions or removals) can be processed and applied to ontologies without having to perform all the reasoning steps from scratch. Pellet has been extended with tableau-based decision procedures for several *E*-Connection[16] languages. Pellet also supports reasoning with SWRL rules. It provides conjunctive query answering and supports SPARQL queries. Pellet reasons over ontologies through Jena and the OWL API, and also supports debugging.

Unlike other DL reasoners, Pellet was designed to work with OWL, so the tableau reasoner has the ability to reason with instance data (ABox reasoning) without making the unique name assumption. Pellet also features an OWL syntax checker, an XML Schema datatype reasoner, and a query engine.

Pellet supports standard inference services, such as consistency checking, concept satisfiability, classification, as well as entailment and conjunctive query answering. Similar to other DL reasoners, Pellet implements optimization techniques, e.g., normalization and simplification, TBox absorption, semantic branching, dependency-directed backjumping, optimized blocking, top-bottom search for classification, and model merging. Pellet also features optimizations not provided by any other DL reasoner, including the following [213]:

---

[15]https://github.com/stardog-union/pellet

[16]A framework for combining several families of decidable logics, such as description logics, modal logic, and some spatial and temporal logics.

- *Nominal absorption.* An improved absorption technique, which absorbs those axioms that declare enumerations into ABox assertions.
- *Partial backjumping.* To prevent useful information from being lost by back-jumping, the dependency set information is inspected to keep useful information and avoid the repeated use of the same tableau rules.
- *Learning-based disjunct selection.* Keeping track of the successful disjunct selections enables the identification of the disjunct that caused less clashes in previous applications, which makes a significant improvement in similar knowledge bases with a large number of individuals.
- *Nominal-based model merging.* Pellet exploits the fixed interpretation of nominals within a domain to improve the model merging algorithm for detecting obvious non-subsumption optimizations for ABox query answering.
- *Optimizations for ABox query answering.* Depending on the structure of the query, different techniques are used, such as rolling-up, which reduces query answering to KB consistency checking.

### 6.6.3   FaCT++

*FaCT++*[17] (Fast Classification of Terminologies) is a tableau-based OWL 2 DL reasoner, with a partial support for OWL 2 key constraints and datatypes. It can be used as a description logic classifier and for modal logic satisfiability testing. It implements a sound and complete tableau algorithm for expressive description logics. FaCT++ is available as a stand-alone tool, as a Protégé plugin, and can be used in applications through the OWL API.

After loading a knowledge base into FaCT++, it is normalized and transformed into an internal representation [214]. During this process, several optimizations are applied, including the following:

- *Lexical normalization and simplification,* a standard rewriting optimization aimed at early and trivial clash detection and concept simplification.
- *Absorption,* which attempts to eliminate GCI axioms to significantly improve reasoning performance.
- *Told cycle elimination,* which transforms axioms to avoid TBox cycles that could lead to expensive subsumption tests by computing a set of (trivially obvious) told subsumers (see Definition 6.10) and told disjoints of concepts (see Definition 6.11).

**Definition 6.10 (Told Subsumer)** Consider a knowledge base $\mathcal{K} = \langle \mathcal{T}, \mathcal{A}, \mathcal{R} \rangle$. If $\mathcal{T}$ contains $A \sqsubseteq C$ or $A = C$, then $C$ is called a *told subsumer* of $A$, denoted as $A \rightarrow_{\text{ts}} C$.

---

[17]http://owl.man.ac.uk/factplusplus/

**Definition 6.11 (Told Disjoint)**  Consider a knowledge base $\mathcal{K} = \langle \mathcal{T}, \mathcal{A}, \mathcal{R} \rangle$. If $\mathcal{T}$ contains an axiom of the form $C \sqsubseteq \neg D \sqcap ...$, then $D$ is called a *told disjoint* of $C$.

- *Synonym replacement*, which further simplifies DL axioms and improves early clash detection.

A core component of FaCT++ is its classifier, which uses a KB satisfiability checker in order to decide subsumption problems for given pairs of concepts, and employs the following optimizations:

- *Dependency-directed backtracking (backjumping)*, involves each concept in a completion tree label being labeled with a dependency set, which contains the branching decisions. When a clash occurs, the system backtracks to the most recent branching point, where a different choice might eliminate the cause of the clash.
- *Boolean constant propagation (BCP)*, is a common simplification based on the inference rule $(\neg C_1,..., \neg C_n, C_1 \sqcup ... \sqcup C_n \sqcup C)/C$ applied to concepts in node labels.
- *Semantic branching*, is another rewriting optimization, which rewrites disjunctions of the form $C \sqcup D$ as $C \sqcup (\neg C \sqcap D)$, thereby ensuring that if choosing $C$ causes a clash, it will not be added to the node label later by a nondeterministic expansion.
- *Heuristics ordering*, in which heuristics can assist in selecting how expansion rules are applied to the disjunctive concepts in a node label to ensure a good candidate expansion. FaCT++ reduces the cost of computing the heuristic function by caching relevant values for each concept.

Classification optimizations reduce the number of subsumption tests needed by reducing the number of comparisons and substituting cheaper but incomplete comparisons wherever possible. Classification optimizations used by FaCT++ include the following:

- *Definitional ordering*, which uses the syntactic structure of TBox axioms to optimize the order in which the taxonomy is computed. This sometimes enables top-down taxonomy computation, thereby avoiding the need to check for subsumees of newly added concepts. The TBox axiom structure can also be used to avoid potentially expensive subsumption tests by computing a set of trivially obvious told subsumers and told disjoints of concepts.
- *Model merging*, which exploits cached partial models in order to perform a relatively cheap but incomplete non-subsumption test.
- *Completely defined concepts*, a technique used for wide, shallow taxonomies. Identifying a significant subset of concepts whose subsumption relationships are completely defined by told subsumptions can eliminate the need for subsumption tests.
- *Clustering*, which adds virtual concepts to the taxonomy to produce a deeper and more uniform structure.

### *6.6.4  Racer*

*Racer*[18] (Renamed ABox and Concept Expression Reasoner) is an open source,[19] server-side reasoner for building ontology-based applications, available through Java and Common Lisp APIs. It is available as a Protégé plugin, and it can connect to the OWL API.

Racer provides not only standard reasoning mechanisms, but also logical abduction. It implements a highly optimized tableau calculus for the $\mathcal{SRIQ}^{(D)}$ description logic. Racer supports the consistency check of RDF data descriptions and OWL 2 ontologies, and can open multiple ontologies simultaneously for ontology merging. It can find implicit subclass relationships induced by the axioms of an ontology and find synonyms for properties, classes, or instances.

Racer can open online and offline ontology files, as well as data retrieved from OWL/RDF documents via SPARQL queries, e.g., from triplestores such as AllegroGraph. It also supports incremental queries. Racer provides a proprietary conjunctive query language, *new Racer Query Language* (nRQL), which supports negation as failure, numeric constraints with reference to attribute values of different individuals, and substring properties between string attributes.

Racer employs FaCT++ optimization techniques and additional optimization for number restrictions and ABoxes. It is optimized for knowledge bases with a very large TBox and a small ABox with typically less than 100 individuals but many different variants, and for knowledge bases with a small TBox and a large ABox to query (more than 100,000 individuals).

## 6.7   Image Interpretation

Description logic representations can utilize domain-specific human knowledge, including expert knowledge, for image interpretation. DL-based image interpretation approaches differ in terms of modeling paradigm (conventional OWA or CWA-like model), reasoning (standard deductive reasoning or abductive reasoning to better model ambiguity), and imprecision handling (probabilistic and fuzzy extensions).

Because of the open world assumption of description logics, incomplete information in image interpretation considers all reasonable explanations, but leaves open which one is the actual situation. For example, by having an axiom stating that landscape images depicting mountains contain at least one region with a mountain, the conventional DL formalism depicts(IMAGE, Mountain), partOf(IMAGE, ROI1), partOf(IMAGE, ROI2) has the following possible interpretations: a

---

[18]http://www.racer-systems.com

[19]https://github.com/ha-mo-we/Racer

mountain is depicted by ROI1, a mountain is depicted by ROI2, or a mountain is depicted by both ROI1 and ROI2. A mountain might also be depicted by other regions of interest (ROIs), and it cannot be assumed that no other ROI depicts a mountain just because there is no information available about it. One way to cope with such open semantics in image interpretation is to incorporate expectation feedback, where the possible interpretations obtained via reasoning provide cues for subsequent analytic cycles when refining the interpretation [215]. Other approaches capture the multiplicity of interpretations by roughly modeling the background knowledge, such as with common sense knowledge bases or TBox axioms defining the typical appearance of snow-capped mountains, upon which a top RoI of the image will very likely depict snow and sky, while a bottom RoI will have meadows, forests, or bare rocks, and definitely not the other way around. In this case, the coverage and sophistication of the formalized background knowledge determines whether the reflection of a mountain in a lake could be interpreted correctly [216], misinterpreted as a mountain, or whether it would lead to contradiction (the sky cannot be below a mountain). Another DL-based modeling paradigm for image interpretation assumes that image analysis provides explicitly asserted descriptions that correspond to all relevant information, which is similar to the closed world assumption.

Deductive reasoning, a standard inference in description logics, can be used for image interpretation. After image segmentation is performed using algorithms such as the Recursive Shortest Spanning Tree (RSST) [217], low-level descriptors of an image (or image sequence) are extracted, which are then analyzed by machine learning algorithms to recognize features, patterns, patches, and regions that correspond to meaningful primitive objects. Some approaches also consider the spatial relation between the depicted objects. The results of feature extraction and analysis are incorporated with background knowledge retrieved from knowledge bases or ontologies. Finally, the task of high-level scene interpretation is realized via reasoning over all these axioms, and potentially identifying meaningful concepts, situations, scenarios, and activities based on the primitive objects provided by low-level analysis and the models of the conceptual knowledge base or ontology (see Fig. 6.3).

Assume the background knowledge about the features of snow-capped mountains, such as shape, dominant colors, concepts frequently depicted together represented as TBox axioms, and a set of facts automatically extracted from landscape images represented as ABox axioms.[20] Deductive reasoning can then verify whether a snow-capped mountain is logically entailed, rather than just identifying rocks or a mountain in general. However, there is a major limitation of deductive reasoning: monotonicity. Once a conclusion is sustained by a valid

---

[20]Often there are serious discrepancies between the intended perceptual to symbolic mappings and the ABox axioms obtained via machine learning. As a consequence, image analysis sometimes fails to provide all the expected descriptions, and generates either false negatives (a depicted concept is not identified) or false positives (concepts are confused with other concepts).

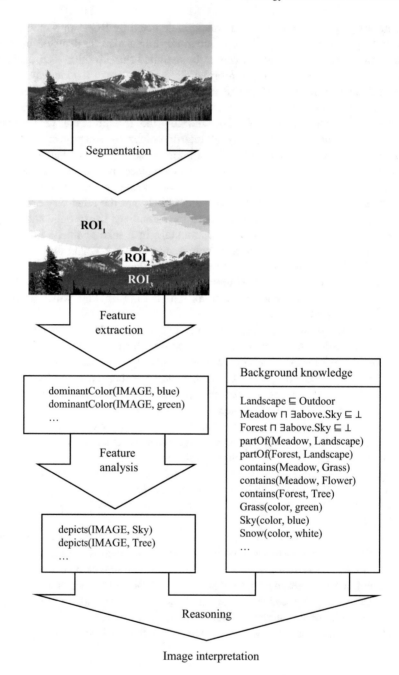

**Fig. 6.3** Architecture of DL-based image interpretation via reasoning

argument, it can never be invalidated, regardless of new assertions added later. In visual knowledge representations, hypotheses are often made using partial evidence by jumping to conclusions, but new information retrieved after making the hypotheses should be able to invalidate the initial hypotheses (*belief revision*). In addition, new individuals cannot be created using deduction. Research efforts aimed at addressing these issues include the combination of deductive reasoning with hypotheses (*hypo-deductive reasoning*) [218], and the inference that seeks the simplest and most likely explanation from a set of potential interpretations while creating a theory from an observation (*abductive reasoning*) [219].

Beyond the incomplete, missing, and contradictory information mentioned above, there is a certain degree of vagueness in image interpretation. The corresponding imprecision can be handled by probabilistic and fuzzy extensions of description logic representations.

### 6.7.1   *Image Interpretation as Abduction*

ABox abduction, which is available in advanced reasoners such as Racer, is suitable for the description logic-based interpretation of information extracted from multimedia resources, such as images and video frames, by providing a symbolic description of the visual content [220]. As an example, interpret the concert scene shown in Fig. 6.4.

Assume knowledge base $\mathcal{K}$ as the input. TBox $\mathcal{T}$ of knowledge base $\mathcal{K}$ covers the music performance domain with DL axioms and DL-safe rules, which are used as background knowledge. The relevant parts of this TBox can be written as follows:

Musician $\sqsubseteq$ Human
partOf(Strings, MusicalIntrument)
partOf(Bow, MusicalInstrument)

**Fig. 6.4** A music performance scene to interpret. Photo by Karl-Heinz Meurer (https://commons.wikimedia.org/wiki/File:Andre_Rieu_2010.jpg)

MusicalPerformance $\sqsubseteq$ $\exists\geqslant$1hasParticipant.Musician
BowedStringInstrumentPerformance $\sqsubseteq$ MusicalPerformance $\sqcap$ $\exists$hasPart.Strings $\sqcap$ $\exists$hasPart.Bow
PluckedStringInstrumentPerformance $\sqsubseteq$ MusicalPerformance $\sqcap$ $\exists$hasPart.Strings
touches(Y,Z) $\leftarrow$ BowedStringInstrumentPerformance(X), hasPart(X, Y ), Bow(Y), hasPart(X,W), Strings(W), hasParticipant(X,Z), Musician(Z)
touches(Y,Z) $\leftarrow$ PluckedStringInstrumentPerformance(X), hasPart(X, Y), Bow(Y), hasParticipant(X,Z), Musician(Z)

Using low-level image analysis, the concepts depicted in the image, along with their spatial relationship, can be formalized in the form of ABox assertions ($\Gamma$) as follows:

Human(HUMAN1)
Strings(STRINGS1)
Bow(BOW1)
touches(HUMAN1, BOW1)

The high-level interpretation of the content in $\Gamma$ can be constructed through abduction by extending the ABox with new concept and role assertions that describe the visual content of the image with richer semantics [221]. The output of the abduction process is formally defined as a set of assertions $E$ such that $\mathcal{K} \cup E \models \Gamma$, where $\mathcal{K} = (\mathcal{T}, \mathcal{A}, \mathcal{R})$ is the knowledge base, $\Gamma$ is a given set of low-level assertions, and $E$ is an explanation to be computed. Explanation $E$ must satisfy the conditions described in Definition 6.8 and can be computed as $\mathcal{K} \cup \Gamma_1 \cup E \models \Gamma_2$, where the assertions in $\Gamma$ will be split into assertions assumed to be true ($\Gamma_1$) and assertions to be explained ($\Gamma_2$), i.e., $\Gamma_1 = \{$Strings (STRINGS1), Human(HUMAN1), Bow(BOW1)$\}$ and $\Gamma_2 = \{$touches (HUMAN1, BOW1)$\}$. The abduction process tries to find explanations ($E$) such that $\Gamma_2$ is entailed. This entailment decision is implemented as a Boolean query answering of the form $Q_1 := \{()|$ touches(HUMAN1, BOW1)$\}$. The output $E$ of the abduction process forms new ABox axioms with multiple potential interpretations, only the following of which are relevant and satisfiable:

E1 = { BowedStringInstrumentPerformance(PERFORMANCE1), hasPart(PERFORMANCE1, BOW1), Strings1(STRINGS1), hasPart(PERFORMANCE1, STRINGS1), Musician(HUMAN1), hasParticipant(PERFORMANCE1, HUMAN1) }

E2 = { BowedStringInstrumentPerformance(PERFORMANCE1), hasPart(PERFORMANCE1, BOW1), hasPart(PERFORMANCE1, STRINGS1), Musician(HUMAN1), hasParticipant(PERFORMANCE1, HUMAN1) }

E3 = { PluckedStringInstrumentPerformance(PERFORMANCE1), hasPart(PERFORMANCE1, BOW1), Musician(HUMAN1), hasParticipant(PERFORMANCE1, HUMAN1) }

The preference measure of the first possible explanation, $E_1$, is calculated by subtracting the number of newly introduced individuals (PERFORMANCE1 and STRINGS1) from the number of individuals considered from $\Gamma_1$ (HUMAN1 and BOW1), i.e., $2 - 2 = 0$. The other values are calculated similarly, which yields to a preference score of 2 for the second possible explanation, $E_2$, and 1 for the third possible explanation, $E_3$. Therefore, the second interpretation is the preferred (most likely) one, suggesting that the image depicts a music performance with a bowed string instrument.

Alternatively, the best explanations of an observation can be calculated through algebraic erosion over the concept lattice of a background theory using *formal concept analysis* (FCA) [222]. Tableau methods are also suitable for generating the preferred hypotheses with reference to a TBox [223].

## 6.7.2 Image Interpretation Using Fuzzy DL Axioms

Fuzzy description logics, such as the f-$\mathcal{SHOIN}$ fuzzy description logic defined in Chap. 4, provide inference support for vague information, which can be utilized in image interpretation tasks, such as object recognition [224]. For example, suppose the following background knowledge of TBox and RBox axioms to be used for reasoning-based image interpretation of a nature photo:

Crown $\sqsubseteq$ color(green) $\sqcap$ texture(patchy)
Trunk $\sqsubseteq$ color(brown) $\sqcap$ texture(rough)
Tree $\equiv$ $\exists$hasPart.(Trunk $\sqcap$ $\exists$isBelow.Crown)
Trans(hasPart)

Assume an image depicting a group of trees segmented, and a set of values produced for each region based on their color and texture. In fuzzy description logics, these values can be described with fuzzy ABox assertions as follows:

color(o1, green) $\geqslant 0.85$
texture(o1, patchy) $\geqslant 0.7$
color(o2, brown) $\geqslant 1.0$
texture(o2, rough) $\geqslant 0.9$
isAbove(o1, o2) $\geqslant 0.9$
hasPart(o3, o2) $\geqslant 0.8$

A fuzzy interpretation $\mathcal{I}$ is a model with reference to the TBox if it holds the following:

$$\text{Crown}^{\mathcal{I}}\left(o_1^{\mathcal{I}}\right) = t\left(\text{color}^{\mathcal{I}}\left(o_1^{\mathcal{I}},\text{green}^{\mathcal{I}}\right),\text{texture}^{\mathcal{I}}\left(o_1^{\mathcal{I}},\text{patchy}^{\mathcal{I}}\right)\right) = t(0.85,0.7)$$
$$\text{Trunk}^{\mathcal{I}}\left(o_2^{\mathcal{I}}\right) = t\left(\text{color}^{\mathcal{I}}\left(o_2^{\mathcal{I}},\text{brown}^{\mathcal{I}}\right),\text{texture}^{\mathcal{I}}\left(o_2^{\mathcal{I}},\text{rough}^{\mathcal{I}}\right)\right) = t(1.0,0.9)$$
$$\text{Tree}^{\mathcal{I}}\left(o_3^{\mathcal{I}}\right) = \sup_b\left\{t\left(\text{hasPart}^{\mathcal{I}}\left(o_3^{\mathcal{I}},b\right),(\text{Trunk}\sqcap\exists\text{isBelow.Crown})^{\mathcal{I}}(b)\right)\right\}$$

$$= \sup_b\left\{t\left(\text{hasPart}^{\mathcal{I}}\left(o_3^{\mathcal{I}},b\right),t\left(\text{Trunk}^{\mathcal{I}}(b),\sup_c\{t(\text{isBelow}^{\mathcal{I}}(b,c),\text{Crown}^{\mathcal{I}}(c))\}\right)\right)\right\}$$

$$\geqslant t\left(\text{hasPart}^{\mathcal{I}}\left(o_3^{\mathcal{I}},o_2^{\mathcal{I}}\right),t\left(\text{Trunk}^{\mathcal{I}}\left(o_2^{\mathcal{I}}\right),t\left((\text{isAbove}^-)^{\mathcal{I}}\left(o_2^{\mathcal{I}},o_1^{\mathcal{I}}\right),\text{Crown}^{\mathcal{I}}\left(o_1^{\mathcal{I}}\right)\right)\right)\right)\geqslant$$

$$t(0.8,t(t(1.0,0.9),t(0.9,t(0.85,0.7))))$$

where $t$ represents the fuzzy intersection performed by a function of the form $t:$ $[0,1]\times[0,1]\rightarrow[0,1]$, called the *t-norm operation*, which must be commutative, i.e., $t(a,b)=t(b,a)$, monotonically increasing, i.e., for $a\leqslant c$ and $b\leqslant d$, $t(a,b)\leqslant t(c,d)$, and associative, i.e., $t(a,t(b,c)) = t(t(a,b),c)$, with 1 being an identity element, i.e., $t(a,1)=a$. Depending on the $t$-norm used, different values can be inferred for $o_3^{\mathcal{I}}$ being a tree. For example, in case of the Łukasiewicz $t$-norm, i.e., $\mathsf{T}_{\text{Luk}}(a,b)=\max\{0,a+b-1\}$, $\text{Tree}^{\mathcal{I}}\left(o_3^{\mathcal{I}}\right)\geqslant 0.15$, the product $t$-norm, i.e., $\mathsf{T}_{\text{prod}}(a,b)=a\cdot b$, gives $\text{Tree}^{\mathcal{I}}\left(o_3^{\mathcal{I}}\right)\geqslant 0.385$, while the minimum $t$-norm (Gödel $t$-norm), i.e., $\mathsf{T}_{\text{min}}(a,b)=\min\{a,b\}$, yields to $\text{Tree}^{\mathcal{I}}\left(o_3^{\mathcal{I}}\right)\geqslant 0.7$.

## 6.8   Video Scene Interpretation

A common approach for ontology-based video understanding is the automated shot annotation with semantic labels using pretrained classifiers. However, frame-level object mapping alone is often insufficient to understand the visual content. To address this limitation, events can be used to provide additional information to interpret a scene. Such information includes object positions, object transitions over time, and the relationship between objects and high-level concepts. However, formalizing complex events by combining primitive events is feasible only for well-defined domains with constrained actions and environment (e.g., soccer videos).

### 6.8.1   Video Event Recognition via Reasoning over Temporal DL Axioms

Using temporal description logics, such as *TL-F*, video events can be formally described and automatically recognized via reasoning. For example, a goal in soccer videos can be described by the sequence goalpost–cheers–closeup–audience–slow motion replay [225] and formalized as follows:

$$\text{GOAL} = \Diamond \left(d_{\text{goal}}, d_{\text{whistle}}, d_{\text{cheers}}, d_{\text{caption}}, d_{\text{goalpost}}, d_{\text{closeup}}, d_{\text{audience}}, d_{\text{replay}}\right) \left(d_{\text{goal}} f\right.$$
$$d_{\text{goalpost}}) \left(d_{\text{whistle}} \, d \, d_{\text{goalpost}}\right) \left(d_{\text{goalpost}} \, o \, d_{\text{cheers}}\right) \left(d_{\text{caption}} \, e \, d_{\text{closeup}}\right) \left(d_{\text{cheers}} \, e \, d_{\text{closeup}}\right) \left(D_{\text{goalpost}}\right.$$
$$m \, d_{\text{closeup}}) \left(d_{\text{closeup}} \, m \, d_{\text{audience}}\right) \left(d_{\text{audience}} \, m \, d_{\text{MSR}}\right) \cdot \left(\text{GOAL} @ d_{\text{goal}} \cap \text{WHISTLE} @\right.$$
$$d_{\text{whistle}} \cap \text{CHEERS} @ d_{\text{cheers}} \cap \text{CAPTION} @ d_{\text{caption}} \cap \text{GOALPOST} @ d_{\text{GOALPOST}} \cap$$
$$\text{CLOSEUP} @ d_{\text{CLOSEUP}} \cap \text{AUDIENCE} @ d_{\text{AUDIENCE}} \cap \text{REPLAY} @ d_{\text{replay}}\right)$$

where $\Diamond$ is the temporal existential quantifier for introducing the temporal intervals; @ is a bindable variable as usual; and $d_{\text{goal}}$, $d_{\text{whistle}}$, $d_{\text{cheers}}$, $d_{\text{caption}}$, $d_{\text{goalpost}}$, $d_{\text{closeup}}$, $d_{\text{audience}}$, and $d_{\text{replay}}$ represent the temporal intervals of the corresponding objects and sequences. After detecting the objects and sequences in a soccer video, they can be described in the form $\Diamond x(\ ).C @ x$, where $C$ is the individual of the object or sequence, $x$ is the temporal interval of $C$, and ( ) denotes those individuals that do not have temporal relationships.

Assume a set of sequence individuals $\{S_0, S_1, \ldots, S_{n-1}, S_n\}$ from the detection results of a soccer video, in which each element $S_i$ can be represented in the form $S_i = \Diamond x_i(\ ).S_i @ x_i$. The definition of $\{S_0, S_1, \ldots, S_{n-1}, S_n\}$ includes a latent temporal constraint, $x_i \, m \, x_{i+1}, i = 0, 1, \ldots, n-1$, which denotes two consecutive sequences in $\{S_0, S_1, \ldots, S_{n-1}, S_n\}$ that are consecutive in the temporal axis of the video. Further assume a set of object individuals $\{O_0, O_1, \ldots, O_{m-1}, O_m\}$ from the detection results of a soccer video, in which each element $O_i$ can be represented in the form $O_i = \Diamond y_i(\ ).O_i @ y_i$.

Based on the above representation, reasoning can be performed over soccer videos to recognize goals as follows. Firstly, those subsets of $\{S_0, S_1, \ldots, S_{n-1}, S_n\}$ are selected that are composed of the consecutive sequence individuals GOALPOST » CLOSEUP » AUDIENCE » REPLAY. These subsets are all goal candidates calculated as $E_{Ck} = \{\text{GOALPOST}_k, \text{CLOSEUP}_{k+1}, \text{AUDIENCE}_{k+2}, \text{REPLAY}_{k+3}\}$, where $k$ is the index of the current view of the current candidate event in $\{S_0, S_1, \ldots, S_{n-1}, S_n\}$. Secondly, all the goal objects $O_{\text{goal}}, O_{\text{whistle}}, O_{\text{cheers}}, O_{\text{caption}}$ have to be found in $\{O_0, O_1, \ldots, O_{m-1}, O_m\}$ for each candidate event $E_{Ck}$, which have a corresponding temporal interval ($y_{\text{goal}}, y_{\text{whistle}}, y_{\text{cheers}}, y_{\text{caption}}$), and satisfy the corresponding temporal constraints, i.e., $y_{\text{goal}} \, f \, \text{GOALPOST}_k, y_{\text{whistle}}$ $d \, \text{GOALPOST}_k, \text{GOALPOST}_k \, o \, y_{\text{cheers}}, y_{\text{caption}} \, e \, \text{CLOSEUP}_{k+1}, y_{\text{cheers}}$ $e \, \text{CLOSEUP}_{k+1}$. If all of such objects exist, $E_{Ck}$ can be considered a goal.

## 6.9 Distributed and Federated Reasoning

Not all ontology-based multimedia processing approaches use a single centralized, consistent ontology for reasoning. *Federated reasoning* is reasoning over data provided by a group of datasets that share a common set of deductive rules; however, this type of semantic reasoning might lead to contradictions

[226]. Asynchronous federated reasoning can be performed to reason with multiple, autonomously developed ontologies, which corresponds to the search for a complete and consistent global tableau [227]. *Distributed reasoning* integrates ontologies and rules with multiple knowledge bases. In multiagent reasoning, where multiple knowledge bases can be assigned to semantic agents, all agents possess and access common taxonomic knowledge as a conceptual hierarchy [228].

## 6.10   Summary

This chapter described how implicit statements can be automatically inferred via reasoning based on the formal representation of concepts depicted in multimedia contents. The most common inference tasks (reasoning services) available for description logics with well-understood computational complexities were presented. It was shown how reasoning enables advanced classification and indexing of multimedia resources, customized data querying, and automated image and video scene interpretation.

# References

1. Yang N-C, Chang W-H, Kuo C-M, Li T-H (2008) A fast MPEG-7 dominant color extraction with new similarity measure for image retrieval. J Vis Comm Image Represent 19(2):92–105. doi:10.1016/j.jvcir.2007.05.003
2. Xu F, Zhang Y-J (2006) Evaluation and comparison of texture descriptors proposed in MPEG-7. J Vis Commun Image Represent 17(4):701–716. doi:10.1016/j.jvcir.2005.10.002
3. Chang C-C, Lin C-J (2011) LIBSVM: a library for support vector machines. ACM Trans Intell Syst Technol 2(3):Article 27. doi:10.1145/1961189.1961199
4. Merler M, Huang B, Xie L, Hua G, Natsev A (2012) Semantic model vectors for complex video event recognition. IEEE Trans Multimed 14(1):88–101. doi:10.1109/TMM.2011. 2168948
5. Luo J, Savakis AE, Singhal A (2005) A Bayesian network-based framework for semantic image understanding. Pattern Recogn 38(6):919–934. doi:10.1016/j.patcog.2004.11.001
6. Papadopoulos GTh, Mezaris V, Kompatsiaris I, Strintzis MG (2007) Combining global and local information for knowledge-assisted image analysis and classification. EURASIP J Adv Signal Process. doi:10.1155/2007/45842
7. Oberhoff D, Kolesnik M (2008) Unsupervised Bayesian network learning for object recognition in image sequences. In: Kůrková V, Neruda R, Koutník J (eds) Artificial neural networks—ICANN 2008. 18th international conference on artificial neural networks, Prague, September 2008, Lecture notes in computer science, vol 5163. Springer, Heidelberg, pp 235–244. doi:10.1007/978-3-540-87536-9_25
8. Viola P, Jones M (2001) Rapid object detection using a boosted cascade of simple features. Paper presented at the 2001 I.E. Computer Society conference on computer vision and pattern recognition, Kauai, 8–14 December 2001, pp 511–518. doi:10.1109/CVPR.2001.990517
9. Lienhart R, Maydt J (2002) An extended set of Haar-like features for rapid object detection. Paper presented at the 2002 international conference on image processing, New York, 22–25 September 2002, pp 900–903. doi:10.1109/ICIP.2002.1038171
10. Lowe DG (2004) Distinctive image features from scale-invariant keypoints. Int J Comput Vis 60(2):91–110. doi:10.1023/B:VISI.0000029664.99615.94
11. Khedher MI, El Yacoubi MA (2015) Local sparse representation-based interest point matching for person re-identification. In: Arik S, Huang T, Lai WK, Liu Q (eds) Neural information processing. 22nd international conference on neural information processing, Istanbul, 9–12 November 2015, Lecture notes in computer science, vol 9491. Springer, Cham, pp 241–250. doi:10.1007/978-3-319-26555-1_28

12. Rublee E, Rabaud V, Konolige K, Bradski G (2011) ORB: an efficient alternative to SIFT or SURF. Paper presented at the 2011 IEEE international conference on computer vision, Barcelona, 6–13 November 2011, pp 2564–2571. doi:10.1109/ICCV.2011.6126544
13. Belhumeur PN, Hespanha JP, Kriegman DJ (1997) Eigenfaces vs. Fisherfaces: recognition using class specific linear projection. IEEE Trans Pattern Anal Mach Intell 19(7):711–720. doi:10.1109/34.598228
14. Kale A, Sundaresan A, Rajagopalan AN, Cuntoor NP, Roy-Chowdhury AK, Kruger V, Chellappa R (2004) Identification of humans using gait. IEEE T Image Process 13 (9):1163–1173. doi:10.1109/TIP.2004.832865
15. Wagner P (2014) Face recognition in videos with OpenCV. http://docs.opencv.org/2.4/ modules/contrib/doc/facerec/tutorial/facerec_video_recognition.html. Accessed 31 Oct 2016
16. Francois A, Nevatia R, Hobbs J, Bolles R, Smith J (2005) VERL: an ontology framework for representing and annotating video events. IEEE Multimed 12(4):76–86. doi:10.1109/MMUL. 2005.87
17. Kennedy L (2006) Revision of LSCOM event/activity annotations. Presented at the DTO challenge workshop on large scale concept ontology for multimedia, Columbia University, New York, December 2006
18. Tani MYK, Lablack A, Ghomari A, Bilasco IM (2015) Events detection using a video surveillance ontology and a rule-based approach. In: Agapito L, Bronstein MM, Rother C (eds) Computer vision—ECCV 2014 workshops. 13th European conference on computer vision, Zürich, September 2014, Lecture notes in computer science, vol 8926. Springer, Cham, pp 299–308. doi:10.1007/978-3-319-16181-5_21
19. Itti L, Koch C, Niebur E (1998) A model of saliency-based visual attention for rapid scene analysis. IEEE Trans Pattern Anal Mach Intell 20(11):1254–1259. doi:10.1109/34.730558
20. Ho C-C, Cheng W-H, Pan T-J, Wu J-L (2003) A user-attention-based focus detection framework and its applications. In: 4th international conference on informations, communications and signal processing and fourth Pacific Rim conference on multimedia, Singapore, December 2003. IEEE, New York, pp 1315–1319. doi:10.1109/ICICS.2003.1292677
21. Ninassi A, Le Meur O, Le Callet P, Barba D, Tirel A (2006) Task impact on the visual attention in subjective image quality assessment. Paper presented at the 14th European signal processing conference, University of Pisa, Florence, 4–8 September 2006
22. Tejero-de-Pablos A, Nakashima Y, Yokoya N, Díaz-Pernas F-J, Martínez-Zarzuela M (2016) Flexible human action recognition in depth video sequences using masked joint trajectories. EURASIP J Image Video Process 2016:20. doi:10.1186/s13640-016-0120-y
23. Chen C, Liu K, Kehtarnavaz N (2016) Real-time human action recognition based on depth motion maps. J Real Time Image Process 12(1):155–163. doi:10.1007/s11554-013-0370-1
24. Lara OD, Labrador MA (2013) A survey on human activity recognition using wearable sensors. IEEE Commun Surv Tutorials 15(3):1192–1209. doi:10.1109/SURV.2012.110112. 00192
25. Chen C, Jafari R, Kehtarnavaz N (2015) A survey of depth and inertial sensor fusion for human action recognition. Multimed Tools Appl 76(3):4405–4425. doi:10.1007/s11042-015-3177-1
26. Nishino K, Kratz L (2013) Modeling crowd flow for video analysis of crowded scenes. In: Ali S, Nishino K, Manocha D, Shah M (eds) Modeling, simulation and visual analysis of crowds, The international series in video computing, vol 11. Springer, New York, pp 237–265. doi:10.1007/978-1-4614-8483-7_10
27. Sikos LF (2014) Image metadata and XMP. In: Web standards: mastering HTML5, CSS3, and XML, 2nd edn. Apress, New York, pp 277–279. doi:10.1007/978-1-4842-0883-0_7
28. Gómez-Romero J, Patricio MA, García J, Molina JM (2010) Ontology-based context representation and reasoning for object tracking and scene interpretation in video. Expert Syst Appl 38(6):7494–7510. doi:10.1016/j.eswa.2010.12.118
29. Grassi M, Morbidoni C, Nucci M (2012) A collaborative video annotation system based on Semantic Web technologies. Cognit Comput 4(4):497–514. doi:10.1007/s12559-012-9172-1

30. Blum A, Mitchell T (1998) Combining labeled and unlabeled data with co-training. Presented at the 11th annual conference on computational learning theory, University of Wisconsin, Madison, 21–24 July 1998

31. Yan R, Liu Y, Jin R, Hauptmann A (2003) On predicting rare class with SVM ensemble in scene classification. In: 2003 IEEE international conference on acoustics, speech and signal processing, Hong Kong, April 2003, vol 3. IEEE, New York, pp 21–24. doi:10.1109/ICASSP. 2003.1199097

32. Hoogs A, Rittscher J, Stein G, Schmiederer J (2003) Video content annotation using visual analysis and large semantic knowledgebase. In: 2003 IEEE Computer Society conference on computer vision and pattern recognition, vol 2. IEEE, New York. doi:10.1109/CVPR.2003. 1211487

33. Naphade M, Smith JR, Tesic J, Chang S-F, Hsu W, Kennedy L, Hauptmann A, Curtis J (2006) Large-scale concept ontology for multimedia. IEEE Multimed 13(3):86–91. doi:10. 1109/MMUL.2006.63

34. Gómez-Romero J, García J, Patricio MA, Serrano MA, Molina JM (2016) Context-based situation recognition in computer vision systems. In: Snidaro L, García J, Llinas J, Blasch E (eds) Context-enhanced information fusion, Advances in computer vision and pattern recognition. Springer, Cham, pp 627–651. doi:10.1007/978-3-319-28971-7_23

35. Hausenblas M, Adida B, Herman I (2008) RDFa—bridging the Web of Documents and the Web of Data. World Wide Web Consortium. https://www.w3.org/2008/Talks/1026-ISCW-RDFa/. Accessed 31 Oct 2016

36. Berners-Lee T (2001) Business model for the Semantic Web. World Wide Web Consortium. https://www.w3.org/DesignIssues/Business. Accessed 31 Oct 2016

37. Powers DMW, Turk CCR (1989) Machine learning of natural language. Springer, London, p 193. doi:10.1007/978-1-4471-1697-4

38. Gruber TR (1993) Towards principles for the design of ontologies used for knowledge sharing. In: Guarino N, Poli R (eds) Formal ontology in conceptual analysis and knowledge representation. Kluwer Academic, Deventer

39. Sikos LF (2015) Structured data. In: Mastering structured data on the Semantic Web: from HTML5 Microdata to Linked Open Data. Apress, New York, pp 2–5. doi:10.1007/978-1-4842-1049-9_1

40. Hayes P (2004) RDFS entailment rules. In: Hayes P (ed) RDF semantics. World Wide Web Consortium. https://www.w3.org/TR/2004/REC-rdf-mt-20040210/#RDFSRules. Accessed 31 Oct 2016

41. Brickley D, Guha RV (2014) RDF Schema 1.1. World Wide Web Consortium. https://www.w3.org/TR/rdf-schema/. Accessed 31 Oct 2016

42. Herman I (2010) Why OWL and not WOL? In: Herman I (ed) Tutorial on Semantic Web technologies. World Wide Web Consortium. https://www.w3.org/People/Ivan/CorePresentations/RDFTutorial/Slides.html#(114). Accessed 31 Oct 2016

43. Dean M, Schreiber G (eds), Bechhofer S, van Harmelen F, Hendler J, Horrocks I, McGuinness DL, Patel-Schneider PF, Stein LA (2004) OWL Web Ontology Language reference. World Wide Web Consortium. https://www.w3.org/TR/owl-ref/. Accessed 31 Oct 2016

44. Hitzler P, Krötzsch M, Parsia B, Patel-Schneider PF, Rudolph S (eds) (2012) OWL 2 Web Ontology Language—Primer, 2nd edn. World Wide Web Consortium. https://www.w3.org/TR/owl-primer/. Accessed 31 Oct 2016

45. Calvanese D, Carroll J, De Giacomo G, Hendler J, Herman I, Parsia B, Patel-Schneider PF, Ruttenberg A, Sattler U, Schneider M (2012) OWL 2 Web Ontology Language—Profiles, 2nd edn. Motik B, Grau BC, Horrocks I, Wu Z, Fokoue A, Lutz C (eds), World Wide Web Consortium. https://www.w3.org/TR/owl2-profiles/. Accessed 16 Apr 2017

46. Miles A, Bechhofer S (eds) (2009) SKOS Simple Knowledge Organization System reference. World Wide Web Recommendation. https://www.w3.org/TR/skos-reference/. Accessed 28 Sept 2016

47. Isaac A, Summers E (eds) Broader/narrower relationships. In: Isaac A, Summers E (eds) SKOS Simple Knowledge Organization System Primer. https://www.w3.org/TR/skos-primer/#sechierarchy. Accessed 29 Sept 2016

48. Hepp M, de Bruijn J (2007) GenTax: a generic methodology for deriving OWL and RDF-S ontologies from hierarchical classifications, thesauri, and inconsistent taxonomies. In: Franconi E, Kifer M, May W (eds) The Semantic Web: research and applications. 4th European Semantic Web conference, Innsbruck, June 2007, Lecture notes in computer science, vol 4519. Springer, Heidelberg, pp 129–144. doi:10.1007/978-3-540-72667-8_11

49. Isaac A, Summers E (2009) SKOS concepts and OWL classes. In: Isaac A, Summers E (eds) SKOS Simple Knowledge Organization System primer. https://www.w3.org/TR/skos-primer/#secskosowl. Accessed 27 Sept 2016

50. Rous B (2012) Major update to ACM's computing classification system. Comm ACM 55 (11):12. doi:10.1145/2366316.2366320

51. Horrocks I, Patel-Schneider PF, Boley H, Tabet S, Grosof B, Dean M (2004) SWRL: a Semantic Web Rule Language combining OWL and RuleML. World Wide Web Consortium. https://www.w3.org/Submission/SWRL/. Accessed 31 Oct 2016

52. Bizer C, Heath T, Berners-Lee T (2009) Linked Data—the story so far. Int J Semant Web Inform Syst 5(3):1–22. doi:10.4018/jswis.2009081901

53. Berners-Lee T (2009) Linked Data–design issues. World Wide Web Consortium. https://www.w3.org/DesignIssues/LinkedData.html. Accessed 31 Oct 2016

54. Adida B, Birbeck M, McCarron S, Herman I (eds) (2015) RDFa Core 1.1—third edition: syntax and processing rules for embedding RDF through attributes. World Wide Web Consortium. https://www.w3.org/TR/rdfa-core/. Accessed 31 Oct 2016

55. Sporny M (ed) (2015) RDFa Lite 1.1—second edition. World Wide Web Consortium. https://www.w3.org/TR/rdfa-lite/. Accessed 31 Oct 2016

56. Adida B, Birbeck M, McCarron S, Herman I (eds) (2012) Completing incomplete triples. In: RDFa Core 1.1. https://www.w3.org/TR/2012/REC-rdfa-core-20120607/#s_Completing_Incomplete_Triples. Accessed 31 Oct 2016

57. Hickson I (ed) (2013) HTML Microdata. World Wide Web Consortium. https://www.w3.org/TR/microdata/. Accessed 31 Oct 2016

58. Sporny M, Longley D, Kellogg G, Lanthaler M, Lindström N (2014) JSON-LD 1.0: a JSON-based serialization for Linked Data. World Wide Web Consortium. https://www.w3.org/TR/json-ld/. Accessed 31 Oct 2016

59. Lowe DG (1999) Object recognition from local scale-invariant features. In: 7th IEEE international conference on computer vision, Kerkyra, September 1999. IEEE, New York, pp 1150–1157. doi:10.1109/ICCV.1999.790410

60. Dollár P, Rabaud V, Cottrell G, Belongie S (2005) Behavior recognition via sparse spatio-temporal features. In: 2005 IEEE international workshop on visual surveillance and performance evaluation of tracking and surveillance, Beijing, October 2005. IEEE, New York, pp 65–72. doi:10.1109/VSPETS.2005.1570899

61. Dalal N, Triggs B (2005) Histograms of oriented gradients for human detection. In: 2005 I.E. Computer Society conference on computer vision and pattern recognition, vol 1, San Diego, June 2005. IEEE Computer Society, Washington, pp 886–893. doi:10.1109/CVPR.2005.177

62. Dalal N, Triggs B, Schmid C (2006) Human detection using oriented histograms of flow and appearance. In: Leonardis A, Bischof H, Pinz A (eds) Computer vision—ECCV 2006. 9th European conference on computer vision, vol 2, Graz, May 2006. Lecture notes in computer science, vol 3952. Springer, Heidelberg, pp 428–441. doi:10.1007/11744047_33

63. Bay H, Ess A, Tuytelaars T, Van Gool L (2008) Speeded-up robust features (SURF). Comput Vis Image Understand 110(3):346–359. doi:10.1016/j.cviu.2007.09.014

64. Duong TH, Nguyen NT, Truong HB, Nguyen VH (2015) A collaborative algorithm for semantic video annotation using a consensus-based social network analysis. Expert Syst Appl 42(1):246–258. doi:10.1016/j.eswa.2014.07.046

65. Guo K, Zhang S (2013) A semantic medical multimedia retrieval approach using ontology information hiding. Comput Math Methods Med 2013:Article ID 407917. doi:10.1155/2013/407917

66. Ballan L, Bertini M, Del Bimbo A, Serra G (2010) Semantic annotation of soccer videos by visual instance clustering and spatial/temporal reasoning in ontologies. Multimed Tools Appl 48(2):313–337. doi:10.1007/s11042-009-0342-4

67. Jiang Y-G, Bhattacharya S, Chang S-F, Shah M (2013) High-level event recognition in unconstrained videos. Int J Multimed Inform Retr 2(2):73–101. doi:10.1007/s13735-012-0024-2

68. Lin J, Wang W (2009) Weakly supervised violence detection in movies with audio and video based co-training. In: Muneesawang P, Wu F, Kumazawa I, Roeksabutr A, Liao M, Tang X (eds) Advances in multimedia information processing—PCM 2009. 10th Pacific Rim conference on multimedia, Bangkok, December 2009, Lecture notes in computer science, vol 5879. Springer, Heidelberg, pp 930–935. doi:10.1007/978-3-642-10467-1_84

69. Gong Y, Wang W, Jiang S, Huang Q, Gao W (2008) Detecting violent scenes in movies by auditory and visual cues. In: Huang Y-MR, Xu C, Cheng K-S, Yang J-FK, Swamy MNS, Li S, Ding J-W (eds) Advances in multimedia information processing—PCM 2008. 9th Pacific Rim conference on multimedia, Tainan, December 2008, Lecture notes in computer science, vol 5353. Springer, Heidelberg, pp 317–326. doi:10.1007/978-3-540-89796-5_33

70. Nam J, Alghoniemy M, Tewfik AH (1998) Audio-visual content-based violent scene characterization. In: 1998 IEEE international conference on image processing, Chicago, October 1998. IEEE Computer Society, Washington, pp 353–357. doi:10.1109/ICIP.1998.723496

71. Ballan L, Bertini M, Del Bimbo A, Serra G (2010) Semantic annotation of soccer videos by visual instance clustering and spatial/temporal reasoning in ontologies. Multimed Tools Appl 48(2):313–337. doi:10.1007/s11042-009-0342-4

72. Saatho C, Scherp A (2009) M3O: The Multimedia Metadata Ontology. Presented at the 10th international workshop of the Multimedia Metadata Community on Semantic Multimedia Database Technologies, Graz, 2 Dec 2009

73. Sikos LF (2015) General, access, and structural metadata. In: Mastering structured data on the Semantic Web: from HTML5 Microdata to Linked Open Data. Apress, New York, p 15. doi:10.1007/978-1-4842-1049-9_2

74. Sikos LF (2014) Image metadata and XMP. In: Web standards: mastering HTML5, CSS3, and XML, 2nd edn. Apress, New York, p 293. doi:10.1007/978-1-4842-0883-0

75. Callet P (2014) 3D reconstruction from 3D cultural heritage models. In: Ioannides M, Quak E (eds) 3D research challenges in cultural heritage, Lecture notes in computer science, vol 8355. Springer, Heidelberg, pp 135–142. doi:10.1007/978-3-662-44630-0_10

76. Sfikas K, Pratikakis I, Koutsoudis A, Savelonas M, Theoharis T (2016) Partial matching of 3D cultural heritage objects using panoramic views. Multimed Tools Appl 75(7):3693–3707. doi:10.1007/s11042-014-2069-0

77. Mallik A, Chaudhury S (2012) Acquisition of multimedia ontology: an application in preservation of cultural heritage. Int J Multimed Inform Retr 1(4):249–262. doi:10.1007/s13735-012-0021-5

78. Brutzman D, Harney J, Blais C (2005) X3D fundamentals. In: Geroimenko V, Chen C (eds) Visualizing information using SVG and X3D. Springer, London, pp 63–84. doi:10.1007/1-84628-084-2_3

79. Petit M, Boccon-Gibod H, Mouton C (2012) Evaluating the X3D schema with Semantic Web tools. In: 17th International conference on 3D web technology, Los Angeles, August 2012. ACM, New York, pp 131–138. doi:10.1145/2338714.2338737

80. Sikos LF (2016) Rich semantics for interactive 3D models of cultural artifacts. In: Garoufallou E, Subirats I, Stellato A, Greenberg J (eds) Metadata and semantics research. 10th international conference on metadata and semantics research, Göttingen, November

2016. Communications in Computer and Information Science, vol 672. Springer, Cham. doi:10.1007/978-3-319-49157-8_14

81. Perperis T, Giannakopoulos T, Makris A, Kosmopoulos DI, Tsekeridou S, Perantonis SJ, Theodoridis S (2011) Multimodal and ontology-based fusion approaches of audio and visual processing for violence detection in movies. Expert Syst Appl 38(11):14102–14116. doi:10.1016/j.eswa.2011.04.219

82. Rodríguez-García MÁ, Colombo-Mendoza LO, Valencia-García R, Lopez-Lorca AA, Beydoun G (2015) Ontology-based music recommender system. In: Omatu S, Malluhi QM, Gonzalez SR, Bocewicz G, Bucciarelli E, Giulioni G, Iqba F (eds) 12th international conference on distributed computing and artificial intelligence, Salamanca, June 2015. Advances in intelligent systems and computing, vol 373. Springer, Cham, pp 39–46. doi:10.1007/978-3-319-19638-1_5

83. Gómez-Romero J, Patricio MA, García J, Molina JM (2011) Ontology-based context representation and reasoning for object tracking and scene interpretation in video. Expert Syst Appl 38(6):7494–7510. doi:10.1016/j.eswa.2010.12.118

84. Sikos LF (2011) Advanced (X)HTML5 metadata and semantics for Web 3.0 videos. DESIDOC J Libr Inform Tech 31(4):247–252. doi:10.14429/djlit.31.4.1105

85. Choudhury S, Breslin JG, Passant A (2009) Enrichment and ranking of the YouTube tag space and integration with the Linked Data Cloud. In: The Semantic Web—ISWC 2009. 8th international Semantic Web conference, Chantilly, October 2009. Lecture notes in computer science, vol 5823. Springer, Heidelberg, pp 747–762. doi:10.1007/978-3-642-04930-9_47

86. Sikos LF, Powers DMW (2015) Knowledge-driven video information retrieval with LOD: from semi-structured to structured video metadata. 8th workshop on exploiting semantic annotations in information retrieval, Melbourne, October 2015. ACM, New York, pp 35–37. doi:10.1145/2810133.2810141

87. Dasiopoulou S, Tzouvaras V, Kompatsiaris I, Strintzis MG (2010) Enquiring MPEG-7-based multimedia ontologies. Multimed Tools Appl 46(2):331–370. doi:10.1007/s11042-009-0387-4

88. Boll S, Klas W, Sheth A (1998) Overview on using metadata to manage multimedia data. In: Sheth A, Klas W (eds) Multimedia data management: using metadata to integrate and apply digital media. McGraw-Hill, New York, p 3

89. Hunter J (2001) Adding multimedia to the Semantic Web—building an MPEG-7 ontology. Presented at the 1st international Semantic Web working symposium, Stanford University, Stanford, 29 July–1 Aug 2001

90. Tsinaraki C, Polydoros P, Christodoulakis S (2004) Integration of OWL ontologies in MPEG-7 and TV-Anytime compliant semantic indexing. In: Persson A, Stirna J (eds) Advanced information systems engineering. 16th international conference on advanced information systems engineering, Riga, June 2004, Lecture notes in computer science, vol 3084. Springer, Heidelberg, pp 398–413. doi:10.1007/978-3-540-25975-6_29

91. Isaac A, Troncy R (2004) Designing and using an audio-visual description core ontology. Paper presented at the workshop on core ontologies in ontology engineering, Northamptonshire, 8 October 2004

92. García R, Celma O (2005) Semantic integration and retrieval of multimedia metadata. Paper presented at the 5th international workshop on knowledge markup and semantic annotation, Galway, 7 November 2005

93. Blöhdorn S, Petridis K, Saathoff C, Simou N, Tzouvaras V, Avrithis Y, Handschuh S, Kompatsiaris Y, Staab S, Strintzis M (2005) Semantic annotation of images and videos for multimedia analysis. In: Gómez-Pérez A, Euzenat J (eds) The Semantic Web: research and applications. Second European Semantic Web conference, Heraklion, May–June 2005, Lecture notes in computer science, vol 3532. Springer, Heidelberg, pp 592–607. doi:10.1007/11431053_40

94. Athanasiadis T, Tzouvaras V, Petridis K, Precioso F, Avrithis Y, Kompatsiaris Y (2005) Using a multimedia ontology infrastructure for semantic annotation of multimedia content. Paper presented at the 5th international workshop on knowledge markup and semantic annotation, Galway, 7 November 2005

95. Dasiopoulou S, Tzouvaras V, Kompatsiaris I, Strintzis M (2009) Capturing MPEG-7 semantics. In: Sicilia M-A, Lytras MD (eds) Metadata and semantics. Springer, Boston, pp 113–122. doi:10.1007/978-0-387-77745-0_11

96. Oberle D et al (2007) DOLCE ergo SUMO: on foundational and domain models in the SmartWeb integrated ontology (SWIntO). J Web Semant Sci Serv Agents World Wide Web 5 (3):156–174. doi:10.1016/j.websem.2007.06.002

97. Horvat M, Bogunović N, Ćosić K (2014) STIMONT: a core ontology for multimedia stimuli description. Multimed Tools Appl 73(3):1103–1127. doi:10.1007/s11042-013-1624-4

98. Suárez-Figueroa MC, Atemezing GA, Corcho O (2013) The landscape of multimedia ontologies in the last decade. Multimed Tools Appl 62(2):377–399. doi:10.1007/s11042-011-0905-z

99. Arndt R, Troncy R, Staab S, Hardman L (2009) COMM: a core ontology for multimedia annotation. In: Staab S, Studer R (eds) Handbook on ontologies. Springer, Heidelberg, pp 403–421. doi:10.1007/978-3-540-92673-3_18

100. Zha Z-J, Mei T, Zheng Y-T, Wang Z, Hua X-S (2012) A comprehensive representation scheme for video semantic ontology and its applications in semantic concept detection. Neurocomputing 95:29–39. doi:10.1016/j.neucom.2011.05.044

101. Jeong JW, Hong HK, Lee DH (2011) Ontology-based automatic video annotation technique in smart TV environment. IEEE Trans Cons Electron 57(4):1830–1836. doi:10.1109/TCE.2011.6131160

102. Hummel B, Thiemann W, Lulcheva I (2008) Scene understanding of urban road intersections with description logic. In: Cohn AG, Hogg DC, Möller R, Neumann B (eds) Dagstuhl seminar proceedings, Dagstuhl, February 2008. Leibniz-Zentrum für Informatik, Dagstuhl, Article # 08091

103. Krötzsch M (2010) Description logic rules. IOS Press, Amsterdam

104. Tao J, Sirin E, Bao J, McGuinness DL (2010) Integrity constraints in OWL. Paper presented at the 24th AAAI conference on artificial intelligence, Westin Peachtree Plaza, Atlanta, 11–15 July 2010

105. Knorr M, Alferes JJ, Hitzler P (2011) Local closed world reasoning with description logics under the well-founded semantics. Artif Intell 175(9–10):1528–1554. doi:10.1016/j.artint.2011.01.007

106. Grau BC, Horrocks I, Botik M, Parsia B, Patel-Schneider P, Sattler U (2008) OWL 2: the next step for OWL. J Web Semant Sci Serv Agents World Wide Web 6(4):309–322. doi:10.1016/j.websem.2008.05.001

107. Horrocks I, Kutz O, Sattler U (2006) The even more irresistible $\mathcal{SROIQ}$. In: Doherty P, Mylopoulos J, Welty CA (eds) Proceedings of the 10th international conference on principles of knowledge representation and reasoning. AAAI Press, Menlo Park, CA, pp 57–67

108. Sikos LF (2015) Mastering structured data on the Semantic Web: from HTML5 Microdata to Linked Open Data. Apress, New York. doi:10.1007/978-1-4842-1049-9

109. Krötzsch M, Simančík F, Horrocks I (2013) A description logic primer. arXiv: https://arxiv.org/abs/1201.4089v3

110. Hustadt U, Motik B, Sattler U (2006) Data complexity of reasoning in very expressive description logics. In: Nineteenth international joint conference on artificial intelligence, Edinburgh, July–August 2005. Morgan Kaufmann, San Francisco, pp 466–471

111. Baader F, Horrocks I, Sattler U (2007) Description logics. In: Van Harmelen F, Lifschitz V, Porter B (eds) Handbook of knowledge representation. Elsevier, Oxford, pp 135–180

112. Reiter R, Mackworth AK (1989) A logical framework for depiction and image interpretation. Artif Intell 41(2):125–155. doi:10.1016/0004-3702(89)90008-8

113. Sofronie-Stokkermans V (2008) Non-classical logics: an introduction. http://people.mpi-inf.mpg.de/~sofronie/seminar-decproc08/slides/slides-modal-description-2-06-08.pdf. Accessed 1 Nov 2016

114. Schmidt-Schauß M, Smolka G (1991) Attributive concept descriptions with complements. Artif Intell 48(1):1–26. doi:10.1016/0004-3702(91)90078-X

115. Kazakov Y, Klinov P (2013) The benefits of incremental reasoning in OWL EL. Paper presented at the 12th international Semantic Web conference, Sydney, 23 October 2013
116. Baader F, Brandt S, Lutz C (2005) Pushing the EL envelope. http://lat.inf.tu-dresden.de/research/reports/2005/BaaderBrandtLutz-LTCS-05-01.ps.gz. Accessed 1 Nov 2016
117. Haarslev V, Lutz C, Möller R (1998) Foundations of spatioterminological reasoning with description logics. In: Cohn AG, Schubert L, Shapiro S (eds) Proceedings of the 6th international conference on principles of knowledge representation and reasoning, Trento, June 1998. Morgan Kaufmann, pp 112–123
118. Haarslev V (1999) A logic-based formalism for reasoning about visual representations. J Vis Lang Comput 4(10):421–445. doi:10.1006/jvlc.1999.0133
119. Haarslev V, Lutz C, Möller R (1999) A description logic with concrete domains and a role-forming predicate operator. J Logic Comput 9:351–384. doi:10.1093/logcom/9.3.351
120. Kaplunova A, Haarslev V, Möller R (2002) Adding ternary complex roles to $\mathcal{ALC}_{RP}^{(D)}$. Paper presented at the international workshop on description logics, Toulouse, 19–21 Apr 2002
121. Randell DA, Cui Z, Cohn AG (1992) A spatial logic based on regions and connections. In: Nebel B, Rich C, Swartout W (eds) Principles of knowledge representation and reasoning. Morgan Kaufmann, pp 165–176
122. Egenhofer MJ (1991) Reasoning about binary topological relations. In: Günther O, Schek H-J (eds) Advances in spatial databases. 2nd symposium on advances in spatial databases, Zürich, August 1991, Lecture notes in computer science, vol 525. Springer, Heidelberg, pp 141–160. doi:10.1007/3-540-54414-3_36
123. Wessel M (2001) Obstacles on the way to qualitative spatial reasoning with description logics: some undecidability results. Paper presented at the 2001 international description logics workshop, Stanford, 1–3 August
124. Na K-S, Kong H, Cho M, Kim P, Baik D-K (2006) Multimedia information retrieval based on spatiotemporal relationships using description logics for the Semantic Web. Int J Intell Syst 21(7):679–692. doi:10.1002/int.20153
125. Wessel M (2003) Qualitative spatial reasoning with the $\mathcal{ALCI}_{RCC}$ family—first results and unanswered questions. https://kogs-www.informatik.uni-hamburg.de/publikationen/pub-wessel/report7.pdf. Accessed 1 Nov 2016
126. Hudelot C, Atif J, Bloch I (2015) $\mathcal{ALC}$(F): a new description logic for spatial reasoning in images. In: Agapito L, Bronstein MM, Rother C (eds) Computer vision—ECCV 2014 workshops. European conference on computer vision, Zürich, September 2014. Lecture notes in computer science, vol 8926. Springer, Cham, pp 370–384. doi:10.1007/978-3-319-16181-5_26
127. Cristani M, Gabrielli N (2011) Practical issues of description logics for spatial reasoning. Paper presented at the AAAI Spring symposium, Stanford University, Stanford, 21–23 Mar 2011
128. Halpern JY, Shoham Y (1991) A propositional modal logic of time intervals. J ACM 38 (4):935–962. doi:10.1145/115234.115351
129. Lutz C (2001) Interval-based temporal reasoning with general TBoxes. In: Nebel B (ed) Proceedings of the 17th international joint conference on artificial intelligence, Seattle, August 2001. Morgan Kaufmann, pp 89–94
130. Artale A, Franconi E (1998) A temporal description logic for reasoning about actions and plans. J Artif Intell Res 9(1):463–506. doi:10.1613/jair.516
131. Bai L, Lao S, Zhang W, Jones GJF, Smeaton AF (2008) Video semantic content analysis framework based on ontology combined MPEG-7. In: Boujemaa N, Detyniecki M, Nürnberger A (eds) Adaptive multimedia retrieval: retrieval, user, and semantics. 5th international workshop on adaptive multimedia retrieval, Paris, July 2007, Lecture notes in computer science, vol 4918. Springer, Heidelberg, pp 237–250. doi:10.1007/978-3-540-79860-6_19
132. Liu W, Xu W, Wang D, Liu Z, Zhang X (2012) A temporal description logic for reasoning about action in event. Inform Tech J 11(9):1211–1218. doi:10.3923/itj.2012.1211.1218
133. Günsel C, Wittmann M (2001) Towards an implementation of the temporal DL TL-$\mathcal{ALC}$. Paper presented at the 2001 international description logics workshop, Stanford University, Stanford, 1–3 Aug 2001

134. Allen JF (1983) Maintaining knowledge about temporal intervals. Comm ACM 26 (11):832–843. doi:10.1145/182.358434

135. Lutz C (2012) A correspondence between temporal description logics. J Appl Non Class Logic 14(1–2):209–233. doi:10.3166/jancl.14.209-233

136. Baader F, Lippmann M (2014) Runtime verification using the temporal description logic $\mathcal{ALC}$-LTL revisited. J Appl Logic 12(4):584–613. doi:10.1016/j.jal.2014.09.001

137. Kamide N (2010) A compatible approach to temporal description logics. Paper presented at the 23rd international workshop on description logics, Waterloo, 4–7 May 2010

138. Gutiérrez-Basulto V, Jung JC, Schneider T (2014) Lightweight temporal description logics with rigid roles and restricted TBoxes. Paper presented at the 24th international joint conference on artificial intelligence, Sheraton Hotel and Convention Center, Buenos Aires, 25–31 July 2015

139. Artale A, Franconi E, Wolter F, Zakharyaschev M (2002) A temporal description logic for reasoning over conceptual schemas and queries. In: Flesca S, Greco S, Ianni G, Leone N (eds) Logics in artificial intelligence. 8th European conference on logics in artificial intelligence, Cosenza, September 2002, Lecture notes in computer science, vol 2424. Springer, Heidelberg, pp 98–110. doi:10.1007/3-540-45757-7_9

140. Hu K, Yu X, Li Z, Zhu H (2010) The temporal description logic TL-$\mathcal{SI}$ and its decidability algorithm. In: 2010 international conference on computational aspects of social networks, Taiyuan, September 2010. IEEE Computer Society, Washington, pp 575–578. doi:10.1109/CASoN.2010.133

141. Artale A, Kontchakov R, Ryzhikov V, Zakharyaschev M (2011) Tailoring temporal description logics for reasoning over temporal conceptual models. In: Tinelli C, Sofronie-Stokkermans V (eds) Frontiers of combining systems. 8th international symposium on frontiers of combining systems, Saarbrücken, October 2011, Lecture notes in computer science, vol 6989. Springer, Heidelberg, pp 1–11. doi:10.1007/978-3-642-24364-6_1

142. Pan D, Zhang Y, Wang D (2012) On formalization of metadata in XBRL based on a temporal description logic $TDL_{BR}^{\star}$. J Inform Comput Sci 9(15):4477–4484

143. Wang Y, Chang L, Li F, Gu T (2014) Verification of branch-time property based on dynamic description logic. In: Shi Z, Wu Z, Leake D, Sattler U (eds) Intelligent information processing. 8th IFIP TC 12 international conference, Hangzhou, October 2014, IFIP advances in information and communication technology, vol 432. Springer, Heidelberg, pp 161–170. doi:10.1007/978-3-662-44980-6_18

144. Milea V, Frasincar F, Kaymak U (2012) tOWL: A temporal web ontology language. IEEE Trans Syst Man Cybern B 42(1):268–281. doi:10.1109/TSMCB.2011.2162582

145. Frasincar F, Milea V, Kaymak U (2010) tOWL: integrating time in OWL. In: de Virgilio R, Giunchiglia F, Tanca L (eds) Semantic Web information management. Springer, Heidelberg, pp 225–246. doi:10.1007/978-3-642-04329-1_11

146. Artale A, Kontchakov R, Ryzhikov V, Zakharyaschev M (2015) Interval temporal description logics. Paper presented at the 28th international workshop on description logics, Athens, 7–10 June 2015

147. Sanati MY (2015) A metric interval-based temporal description logic. Ph.D. thesis, McMaster University, Hamilton, June 2015

148. Sotnykova A, Vangenot C, Cullot N, Bennacer N, Aufaure M-A (2005) Semantic mappings in description logics for spatio-temporal database schema integration. In: Spaccapietra S, Zimányi E (eds) Journal on data semantics III, Lecture notes in computer science, vol 3534. Springer, Heidelberg, pp 143–167. doi:10.1007/11496168_7

149. Elleuch N, Zarka M, Ammar AB, Alimi AM (2011) A fuzzy ontology-based framework for reasoning in visual video content analysis and indexing. In: Eleventh international workshop on multimedia data mining, Manchester Grand Hyatt, San Diego, 21–24 Aug 2011. doi:10.1145/2237827.2237828

150. Liu J, Cui LL (2014) Extending description logic $\mathcal{SHOIN}$ based on cloud model. Appl Mech Mater 543–547:3586–3589. doi:10.4028/www.scientific.net/AMM.543-547.3586

151. Stoilos G, Stamou G, Tzouvaras V, Pan J, Horrocks I (2005) The fuzzy description logic f-$\mathcal{SHIN}$. Paper presented at the ISWC workshop on uncertainty reasoning for the Semantic Web, Galway, 7 Nov 2005

152. Stoilos G, Stamou G, Pan JZ (2010) Fuzzy extensions of OWL: logical properties and reduction to fuzzy description logics. Int J Approx Reason 51(6):656–679. doi:10.1016/j.ijar.2010.01.005

153. Grosof BN, Horrocks I, Volz R, Decker S (2003) Description logic programs: combining logic programs with description logic. In: 12th international conference on World Wide Web. ACM, New York, pp 48–57. doi:10.1145/775152.775160

154. Krötzsch M, Rudolph S, Hitzler P (2008) Description logic rules. In: Ghallab M, Spyropoulos CD, Fakotakis N, Avouris N (eds) 18th European conference on artificial intelligence. 18th European conference on artificial intelligence, Patras, February 2008, Frontiers in artificial intelligence and applications, vol 178. IOS Press, Amsterdam, pp 80–84. doi:10.3233/978-1-58603-891-5-80

155. Sikos LF (2017) 3D model indexing in videos for content-based retrieval via X3D-based semantic enrichment and automated reasoning. In: Invited paper. 22nd international conference on 3D web technology, Brisbane, June 2017. ACM, New York. doi:10.1145/3055624.3075943

156. Cano P, Batlle E, Kalker T, Haitsma J (2005) A review of audio fingerprinting. J VLSI Sign Process Syst Sign Image Video Technol 41(3):271–284. doi:10.1007/s11265-005-4151-3

157. Sikos LF (2016) RDF-powered semantic video annotation tools with concept mapping to Linked Data for next-generation video indexing: a comprehensive review. Multimed Tool Appl. doi:10.1007/s11042-016-3705-7

158. Petković M, Jonker W (2004) Spatio-temporal formalization of video events. In: Content-based video retrieval, The Springer international series in engineering and computer science, vol 25. Springer, New York, pp 55–71. doi:10.1007/978-1-4757-4865-9_4

159. Ronfard R (1997) Shot-level description and matching of video content. In: Jay Kuo C-C, Chang S-F, Gudivada VN (eds) Multimedia storage and archiving systems II. The International Society for Optical Engineering, Bellingham. doi:10.1117/12.290366

160. Welty C, Fikes R (2006) A reusable ontology for fluents in OWL. In: Proceedings of the 4th international conference on formal ontology in information systems. IOS, Amsterdam, pp 226–236

161. Hayes P, Welty P (2006) Defining n-ary relations on the Semantic Web. Noy N, Rector A (eds), https://www.w3.org/TR/swbp-n-aryRelations/. Accessed 3 Nov 2016

162. Batsakis S, Petrakis EGM (2011) SOWL: a framework for handling spatio-temporal information in OWL 2.0. In: Bassiliades N, Governatori G, Paschke A (eds) Rule-based reasoning, programming, and applications. 5th international conference on rule-based reasoning, programming, and applications, Barcelona, July 2011, Lecture notes in computer science, vol 6826. Springer, Heidelberg, pp 242–249. doi:10.1007/978-3-642-22546-8_19

163. Perperis T, Giannakopoulos T, Makris A, Kosmopoulos DI, Tsekeridou S, Perantonis SJ, Theodoridis S (2011) Multimodal and ontology-based fusion approaches of audio and visual processing for violence detection in movies. Expert Syst Appl 38(11):14102–14116. doi:10.1016/j.eswa.2011.04.219

164. Sikos LF (2015) Interlinking. In: Mastering structured data on the Semantic Web: from HTML5 Microdata to Linked Open Data. Apress, New York, pp 72–74. doi:10.1007/978-1-4842-1049-9_3

165. Noy NF, McGuinness DL (2001) Ontology development 101: a guide to creating your first ontology. http://protege.stanford.edu/publications/ontology_development/ontology101.pdf. Accessed 3 Nov 2016
166. Sikos LF (2015) RDB to RDF direct mapping. In: Mastering structured data on the Semantic Web: from HTML5 Microdata to Linked Open Data. Apress, New York, pp 217–220. doi:10.1007/978-1-4842-1049-9_9
167. Cranefield S, Purvis M (1999) UML as an ontology modelling language. Paper presented at the workshop on intelligent information integration, 16th international joint conference on artificial intelligence, City Conference Center, Stockholm, 31 July–6 Aug 1999
168. Pinet F, Roussey C, Brun T, Vigier F (2009) The use of UML as a tool for the formalisation of standards and the design of ontologies in agriculture. In: Pardalos PM, Papajorgji PJ (eds) Advances in modeling agricultural systems, Springer optimization and its applications, vol 25. Springer, New York, pp 131–147. doi:10.1007/978-0-387-75181-8_7
169. Sikos LF (2016) A novel approach to multimedia ontology engineering for automated reasoning over audiovisual LOD datasets. In: Nguyen NT, Trawiński B, Fujita H, Hong T-P (eds) Intelligent information and database systems. 8th Asian conference on intelligent information and database systems, Đà Nẵng, March 2016, Lecture notes in computer science (Lecture notes in artificial intelligence), vol 9621. Springer, Heidelberg, pp 3–12. doi:10.1007/978-3-662-49381-6_1
170. Simperl E (2009) Reusing ontologies on the Semantic Web: a feasibility study. Data Knowl Eng 68(10):905–925. doi:10.1016/j.datak.2009.02.002
171. Gandon F (2002) Distributed artificial intelligence and knowledge management: ontologies and multiagent systems for a corporate Semantic Web. Ph.D. thesis, University of Nice, Nice, November 2002
172. Gruber TR (1995) Towards principles for the design of ontologies used for knowledge sharing? Int J Hum Comput Stud 43(5–6):907–928. doi:10.1006/ijhc.1995.1081
173. Lange C, Kohlhase M (2009) A mathematical approach to ontology authoring and documentation. In: Carette J, Dixon L, Coen CS, Watt SM (eds) Intelligent computer mathematics. 16th symposium on the integration of symbolic computation and mechanised reasoning, Grand Bend, July 2009, Lecture notes in computer science, vol 5625. Springer, Heidelberg, pp 389–404. doi:10.1007/978-3-642-02614-0_31
174. Altova (2012) Altova SemanticWorks 2012 user and reference manual. https://www.altova.com/documents/SemanticWorks.pdf. Accessed 4 Nov 2016
175. TopQuadrant (2015) TopBraid Composer Standard Edition. http://www.topquadrant.com/tools/modeling-topbraid-composer-standard-edition/. Accessed 16 Apr 2017
176. TopQuadrant (2015) TopBraid Composer Maestro Edition. http://www.topquadrant.com/tools/ide-topbraid-composer-maestro-edition/. Accessed 16 Apr 2017
177. Sikos LF, Powers DMW (2015) Knowledge-driven video information retrieval with LOD: from semi-structured to structured video metadata. In 8th workshop on exploiting semantic annotations in information retrieval, Melbourne, October 2015. ACM, New York, pp 35–37. doi: 10.1145/2810133.2810141
178. Simou N, Tzouvaras V, Avrithis Y, Stamou G, Kollias S (2005) A visual descriptor ontology for multimedia reasoning. Paper presented at the 6th international workshop on image analysis for multimedia interactive services, Montreux, 13–15 Apr 2005
179. Alexander A, Meehleib T (2001) The thesaurus for graphic materials: its history, use, and future. Cat Classif Q 31(3–4):189–212. doi:10.1300/J104v31n03_04
180. Saathoff C, Timmermann N, Staab S, Petridis K, Anastasopoulos D, Kompatsiaris Y (2006) M-OntoMat-Annotizer: linking ontologies with multimedia low-level features for automatic image annotation. Paper presented at the 3rd European Semantic Web conference, Budva, 11–14 June 2006
181. Morbidoni C, Piccioli A (2015) Curating a document collection via crowdsourcing with Pundit 2.0. In: Gandon F, Guéret C, Villata S, Breslin J, Faron-Zucker C, Zimmermann A (eds) The Semantic Web: ESWC 2015 satellite events. ESWC 2015, Portorož, May–June

2015, Lecture notes in computer science, vol 9341. Springer, Cham, pp 102–106. doi:10. 1007/978-3-319-25639-9_20

182. Iakovidis DK, Goudas T, Smailis C, Maglogiannis I (2014) Ratsnake: a versatile image annotation tool with application to computer-aided diagnosis. Sci World J. Hindawi, Cairo, Article ID 286856. doi:10.1155/2014/286856

183. Restagno L, Akkermans V, Rizzo G, Servetti A (2011) A semantic web annotation tool for a web-based audio sequencer. In: Auer S, Díaz O, Papadopoulos GA (eds) Web engineering. 11th international conference on web engineering, Paphos, June 2011, Lecture notes in computer science, vol 6757. Springer, Heidelberg, pp 289–303. doi:10.1007/978-3-642-22233-7_20

184. Sikos LF (2016) RDF-powered semantic video annotation tools with concept mapping to Linked Data for next-generation video indexing: a comprehensive review. Multimed Tool Appl. doi:10.1007/s11042-016-3705-7

185. Oomoto E, Tanaka K (1993) OVID: design and implementation of a video-object database system. IEEE Trans Knowl Data Eng 5(4):629–643. doi:10.1109/69.234775

186. Carrer M, Ligresti L, Ahanger G, Little TDC (1998) An annotation engine for supporting video database population. In: Furht B (ed) Multimedia technologies and applications for the 21st century, The Springer international series in engineering and computer science, vol 431. Springer, New York, pp 161–184. doi:10.1007/978-0-585-28767-6_7

187. Aydınlılar M, Yazıcı A (2013) Semi-automatic semantic video annotation tool. In: Gelenbe E, Lent R (eds) Computer and information sciences III, 27th international symposium on computer and information sciences, Paris, October 2012. Springer, London, pp 303–310. doi:10.1007/978-1-4471-4594-3_31

188. Bellini P, Nesi P, Serena M (2015) MyStoryPlayer: experiencing multiple audio-visual content for education and training. Multimed Tool Appl 74(18):8219–8259. doi:10.1007/s11042-014-2052-9

189. Sikos LF, Powers DMW (2015) Knowledge-driven video information retrieval with LOD: from semi-structured to structured video metadata. In: 8th workshop on exploiting semantic annotations in information retrieval, Melbourne, October 2015. ACM, New York, pp 35–37. doi:10.1145/2810133.2810141

190. Pittarello F, De Faveri A (2006) Semantic description of 3D environments: a proposal based on web standards. In: 11th international conference on 3D web technology, Columbia, April 2006. ACM, New York, pp 85–95. doi:10.1145/1122591.1122603

191. Bilasco IM, Gensel J, Villanova-Oliver M, Martin H (2006) An MPEG-7 framework enhancing the reuse of 3D models. In: 11th international conference on 3D web technology, Columbia, April 2006. ACM, New York, pp 65–74. doi:10.1145/1122591.1122601

192. Havemann S, Settgast V, Berndt R, Eide Ø, Fellner DW The Arrigo showcase reloaded— towards a sustainable link between 3D and semantics. J Comput Cult Herit 2(1):Article 4. doi:10.1145/1551676.1551680

193. Serna SP, Scopigno R, Doerr M, Theodoridou M, Georgis C, Ponchio F, Stork A (2011) 3D-centered media linking and semantic enrichment through integrated searching, browsing, viewing and annotating. In: Niccolucci F, Dellepiane M, Serna SP, Rushmeier H, Van Gool L (eds) VAST: international symposium on virtual reality, archaeology and intelligent cultural heritage. 12th international symposium on virtual reality, archaeology and cultural heritage, Prato, October 2011. Eurographics, pp 89–96. doi:10.2312/VAST/VAST11/089-096

194. Rodriguez-Echavarria K, Theodoridou M, Georgis C, Arnold DB, Doerr M, Stork A, Serna SP (2012) Semantically rich 3D documentation for the preservation of tangible heritage. In: Arnold D, Kaminski J, Niccolucci F, Stork A (eds) VAST: international symposium on virtual reality, archaeology and intelligent cultural heritage. 13th international symposium on virtual reality, archaeology and cultural heritage, Brighton, November 2012. Eurographics, pp 41–48. doi:10.2312/VAST/VAST12/041-048

195. Yu D, Hunter J (2014) X3D fragment identifiers—extending the Open Annotation model to support semantic annotation of 3D cultural heritage objects over the Web. Int J Herit Digit Era 3(3):579–596. doi:10.1260/2047-4970.3.3.579

196. Yu D, Hunter J (2014) X3D fragment identifiers—extending the Open Annotation model to support semantic annotation of 3D cultural heritage objects over the Web. Int J Herit Digit Era 3(3):579–596. doi:10.1260/2047-4970.3.3.579

197. Bail S (2013) Common reasons for ontology inconsistency. http://ontogenesis.knowledgeblog.org/1343. Accessed 3 Nov 2016

198. Calvanese D, Giacomo GD, Lembo D, Lenzerini M, Rosati R (2005) DL-Lite: tractable description logics for ontologies. In: 20th national conference on artificial intelligence, vol 2, Pittsburgh, July 2005. AAAI Press, pp 602–607

199. Möller R, Neumann B (2008) Ontology-based reasoning techniques for multimedia interpretation and retrieval. In: Kompatsiaris Y, Hobson P (eds) Semantic multimedia and ontologies: theory and applications. Springer, London, pp 55–98. doi:10.1007/978-1-84800-076-6_3

200. Carral D, Feier C, Grau BC, Hitzler P, Horrocks I (2014) Pushing the boundaries of tractable ontology reasoning. In: Mika P, Tudorache T, Bernstein A, Welty C, Knoblock C, Vrandečić D, Groth P, Noy N, Janowicz K, Goble C (eds) The Semantic Web—ISWC 2014. 13th international Semantic Web conference, Riva del Garda, October 2014. Springer, Cham, pp 148–163. doi:10.1007/978-3-319-11915-1_10

201. Hayes P (2014) RDFS entailment rules. In: Hayes P (ed) RDF semantics. World Wide Web Consortium. https://www.w3.org/TR/2004/REC-rdf-mt-20040210/#RDFSRules. Accessed 31 Oct 2016

202. Ter Horst HJ (2005) Completeness, decidability and complexity of entailment for RDF Schema and a semantic extension involving the OWL vocabulary. J Web Semant Sci Serv Agents World Wide Web 3(2–3):79–115. doi:10.1016/j.websem.2005.06.001

203. Carroll J, Herman I, Patel-Schneider PF (2012) OWL 2 Web Ontology Language RDF-based semantics (2nd edn). Schneider M (ed). https://www.w3.org/TR/owl2-rdf-based-semantics/. Accessed 16 Apr 2017

204. Motik B, Grau BC, Horrocks I, Wu Z, Fokoue A, Lutz C (eds) (2012) Reasoning in OWL 2 RL and RDF graphs using rules. In: OWL 2 Web Ontology Language profiles. https://www.w3.org/TR/owl2-profiles/#Reasoning_in_OWL_2_RL_and_RDF_Graphs_using_Rules. Accessed 16 Apr 2017

205. Motik B, Shearer R, Horrocks I (2009) Hypertableau reasoning for description logics. J Artif Intell Res 36:165–228

206. Glimm B, Horrocks I, Motik B, Stoilos G, Wang Z (2014) HermiT: an OWL 2 reasoner. J Automat Reason 53(3):245–269. doi:10.1007/s10817-014-9305-1

207. Borgwardt S, Peñaloza R (2016) Reasoning in fuzzy description logics using automata. Fuzzy Set Syst 298:22–43. doi:10.1016/j.fss.2015.07.013

208. Carbotta D (2010) A practical automata-based technique for reasoning in expressive description logics. M.Sc. thesis, Technische Universität Wien, Vienna, October 2010

209. Georgieva L, Hustadt U, Schmidt RA (2002) A new clausal class decidable by hyperresolution. In: Voronkov A (ed) Automated deduction—CADE-18. 18th international conference on automated deduction, Copenhagen, July 2002. Lecture notes in computer science, vol 2392. Springer, Heidelberg, pp 260–274. doi:10.1007/3-540-45620-1_21

210. Franconi E, Ibáñez-García YA, Seylan I (2011) Query answering with DBoxes is hard. Electron Notes Theor Comput Sci 278:71–84. doi:10.1016/j.entcs.2011.10.007

211. Dolby J, Fokoue A, Kalyanpur A, Schonberg E, Srinivas K (2009) Scalable highly expressive reasoner (SHER). J Web Semant Sci Serv Agents World Wide Web 7(4):357–361. doi:10.1016/j.websem.2009.05.002

212. Shearer R, Motik B, Horrocks I (2008) HermiT: a highly-efficient OWL reasoner. Paper presented at the 5th OWLED workshop on OWL: experiences and directions, Karlsruhe, 26–27 Oct 2008

213. Sirin E, Parsia B, Grau BC, Kalyanpur A, Katz Y (2007) Pellet: a practical OWL-DL reasoner. J Web Semant Sci Serv Agents World Wide Web 5(2):51–53. doi:10.1016/j.websem.2007.03.004

214. Tsarkov D, Horrocks I (2006) FaCT++ description logic reasoner: system description. In: Furbach U, Shankar N (eds) Automated reasoning. 3rd international joint conference on automated reasoning, Seattle, August 2006. Lecture notes in computer science, vol 4130. Springer, Heidelberg, pp 292–297. doi:10.1007/11814771_26

215. Hotz L, Neumann B, Terzic K (2008) High-level expectations for low-level image processing. In: Dengel AR, Berns K, Breuel TM, Bomarius F, Roth-Berghofer TR (eds) KI 2008: advances in artificial intelligence. 31st annual German conference on AI, Kaiserslautern, September 2008, Lecture notes in computer science (Lecture notes in artificial intelligence), vol 5243. Springer, Heidelberg, pp 87–94. doi:10.1007/978-3-540-85845-4_11

216. Gómez-Romero J, Patricio MA, García J, Molina JM (2011) Ontology-based context representation and reasoning for object tracking and scene interpretation in video. Expert Syst Appl 38(6):7494–7510. doi:10.1016/j.eswa.2010.12.118

217. Kwok SH, Constantinides AG (1997) A fast recursive shortest spanning tree for image segmentation and edge detection. IEEE Trans Image Process 6(2):328–332. doi:10.1109/83.551705

218. Hummel B, Thiemann W, Lulcheva I (2007) Description logic for vision-based intersection understanding. In: Cognitive systems with interactive sensors, Stanford University, Stanford

219. Elsenbroich C, Kutz O Sattler U (2006) A case for abductive reasoning over ontologies. Paper presented at OWL: experiences and directions, Athens, 10–11 Nov 2006

220. Möller R, Neumann B (2008) Ontology-based reasoning techniques for multimedia interpretation and retrieval. In: Kompatsiaris Y, Hobson P (eds) Semantic multimedia and ontologies: theory and application. Springer, London, pp 55–98. doi:10.1007/978-1-84800-076-6_3

221. Peraldi SE, Kaya A, Melzer S, Möller R, Wessel M (2007) Multimedia interpretation as abduction. Paper presented at the 20th international workshop on description logics, Brixen-Bressanone, 8–10 June 2007

222. Atif J, Hudelot C, Bloch I (2013) Explanatory reasoning for image understanding using formal concept analysis and description logics. IEEE Trans Syst Man Cybern A 44 (5):552–570. doi:10.1109/TSMC.2013.2280440

223. Yang Y, Atif J, Bloch I (2015) Abductive reasoning using tableau methods for high-level image interpretation. In: Hölldobler S, Krötzsch M, Peñaloza R, Rudolph S (eds) 38th annual German conference on AI, Dresden, September 2015, Lecture notes in computer science (Lecture notes in artificial intelligence), vol 9324. Springer, Cham, pp 356–365. doi:10.1007/978-3-319-24489-1_34

224. Stoilos G, Stamou G, Pan JZ (2010) Fuzzy extensions of OWL: logical properties and reduction to fuzzy description logics. Int J Approx Reas 51(6):656–679. doi:10.1016/j.ijar.2010.01.005

225. Bai L, Lao S, Zhang W, Jones GJF, Smeaton AF (2008) Video semantic content analysis framework based on ontology combined MPEG-7. In: Boujemaa N, Detyniecki M, Nürnberger A (eds) Adaptive multimedia retrieval: retrieval, user, and semantics. 5th international workshop on adaptive multimedia retrieval, Paris, July 2007, Lecture notes in computer science, vol 4918. Springer, Heidelberg, pp 237–250. doi:10.1007/978-3-540-79860-6_19

226. Cholvy L (1998) Reasoning about data provided by federated deductive databases. J Intell Inform Syst 10(1):49–80. doi:10.1023/A:1008637507908

227. Bao J, Caragea D, Honavar V (2006) A tableau-based federated reasoning algorithm for modular ontologies. In: 2006 IEEE/WIC/ACM international conference on web intelligence, December 2006. IEEE, pp 404–410. doi:10.1109/WI.2006.28

228. Kaneiwa K, Mizoguchi R (2009) Distributed reasoning with ontologies and rules in order-sorted logic programming. J Web Semant Sci Serv Agents World Wide Web 7(3):252–270. doi:10.1016/j.websem.2009.05.003

# Index

## A

Abductive reasoning, 103, 168, 169, 178, 181, 182
ABox, 78–81, 84, 85, 91, 93, 101, 104, 126, 153–155, 169, 170, 172, 174–176, 178, 179, 181–183
Abstract domain. *See* Object domain
Acoustic fingerprint, 110
Acyclic TBox, 80
AKTive Media, 143
$\mathcal{ALC}$, 172
$\mathcal{ALC}$(C), 70, 71, 83, 84, 87, 95–97, 102
$\mathcal{ALC}$(CDC), 90, 91
$\mathcal{ALC}$($D_{RCC8}$), 89
$\mathcal{ALC}$(F), 90
$\mathcal{ALCF}$, 93–96
$\mathcal{ALC}$-LTL, 95–97
$\mathcal{ALC}_{RA\ominus}$, 89
$\mathcal{ALCRP}(S_2\oplus T)$, 102
$\mathcal{ALCRP}^{(\mathcal{D})}$, 87–89, 102
$\mathcal{ALCHIQ}^{(\mathcal{D})}$, 71
Allen's temporal relations, 93, 94, 96, 101, 102, 115
Anvil, 145
Apache Jena, 131, 133, 173, 175
Apache Maven, 134
Apache Stanbol, 132, 133
Atomic role, 13, 14
Audio descriptor, 53
Audio Effects Ontology, 63, 141
Audio Features Ontology, 62, 112
AutoDesk 3ds Max, 108, 109
Automaton, 169, 171

## B

Bag of words, 3
$B\mathcal{ALC}_l$, 97, 98
Bayesian network, 4, 5
BBC Programmes Ontology, 141
Belief revision, 181
Big Data, 34, 45
Bnode, 15, 46, 158

## C

Cardinal Direction Calculus, 90
Cardinality restriction, 27
Cepstral audio descriptor, 54
CGI, 113
Chord Ontology, 62, 141
Classical Music Navigator Ontology, 63
Closed caption, 3
Closed predicate, 173
Closed world assumption, 7, 68, 179
CMSY10 font, 69, 91, 103
Collaborative annotation, 8, 56
Color descriptor, 52, 138
Common sense ontology, 2, 7, 9, 122, 135, 142
Completion graph, 169
Computational complexity, 7, 21, 64, 83, 84, 98, 135, 151, 171–173
Computational Tree Logic, 92, 98
Conceptualization, 7, 9, 11, 35, 78, 112, 123, 125, 127, 137, 140
Concrete domain, 71, 87, 102
Conjunctive query, 85
Content-based video retrieval, 5, 112

Printed in the United States
By Bookmasters